# Fleet Purchasing Maintenance and Reliability

Diarmaid Murphy

A full catalog record for this book is available
from the Library of Congress.

ISBN 978-0-8311-3504-1

Industrial Press, Inc.
32 Haviland Street
South Norwalk, Connecticut 06854

Sponsoring Editor: Jim Dodd
Developmental Editor: Robert Weinstein
Interior Text and Cover Design: Janet Romano-Murray

# Table of Contents

## PART 2 MAINTENANCE SYSTEM

# INTRODUCTION

Although there are many detailed books covering specific and complex individual aspects of fleet management and maintenance, this book aims to address the key fundamentals of purchasing vehicles or fleets, and their subsequent maintenance, in a practical and straightforward manner. I hope to outline the various facets of the numerous topics related to these subjects in a more approachable and manageable way than many readers will have seen before. I will discuss initial vehicle purchasing, operational fleet reliability, life cycle costs, and personnel matters among other topics.

My experience is based primarily on commercial and peacetime military maintenance activities with small-to-medium fleet sizes, but the fundamental ideas discussed can be cascaded upwards or downwards as required. Although many examples used are military oriented, most of the principles and processes apply equally to military and civilian vehicle fleets alike. Furthermore, a method of improving the fundamental performance of a battle tank in strenuous conditions from World War II to the present day will likely improve the performance of a 21st century road transport vehicle.

# Introduction

Whatever name is given to your maintenance system, the procedures in place should not detract from the overall aim which every maintenance system — in every organization large or small – should have in common. This goal is to continuously improve the systems in place, to improve maintenance quality and efficiency, thereby reducing wastage and costs. We will approach the key fleet considerations in a manner that will give non-technical managers a general appreciation of where their fleet maintenance management should be, or where it may have to go, in order to achieve cost effectiveness and the aims of your organization.

> *Many companies are still not aware how inefficient their own maintenance is, or if they are aware, they might not want anyone else to know.*
>
> *(Borris, 2006)*

Throughout this book I will be giving practical guidance on the best ways to manage fleet maintenance and review possible failings in the maintenance management of larger fleets. I do not maintain that this approach is "best practice" because I have a very low tolerance for that term. What is best practice for your individual business or organization are the procedures that give the best return in your unique set of circumstances. Despite an entire industry having grown around marketing the idea of generalized best practice, there are few generic best practices applicable to numerous individual companies across one or more business types. Pursuing generic ideas of best practice can often lead to increased costs and inefficiencies in organizations that are not suited to the implementation of a generic formula.

I do give a cursory mention to a number of the models and theories that populate current thinking regarding efficient vehicle maintenance, but these are to highlight matters of importance and not simply for their own benefit. Likewise, I mention life cycle costs, but this is not a book specifically about that topic. In fact, life cycle costs, as is explained later in the book, is fundamentally an umbrella term and not an individually quantifiable expense.

The costs involved with mismanagement of a fleet of vehicles increase significantly with the size of the fleet. In addition, most large modern organizations will suffer from some level of bureaucracy, red tape, and missed opportunities to save money. These negative aspects of modern business tend to increase dramatically and disproportionately to fleet or organization size. No different from other costs incurred by any large organization, the larger your fleet, the more means there are by which to lose or waste money, and the more inefficiencies can creep in.

Misplaced priorities in fleet maintenance or budget allocation also tend to increase out of proportion to the size of an organization. As the maintenance management chain becomes longer, benefits can be achieved through devolved budgets and localized management. Unfortunately the increase in bureaucracy in larger companies can compound any problems of priorities and finances being misplaced. An increase in maintenance flexibility can often be given by devolved budgets with one hand and taken away by bureaucracy with the other hand. The length of a maintenance chain, with devolved finances, can therefore be a double-edged sword.

The fleet maintenance role is becoming ever more specialized due to the nature of modern vehicles and, indeed, companies. This book is intended not only to put into words what maintenance managers may not realize they already know, but also to provide insight and perspective to senior managers who have overall responsibility for fleets and their associated maintenance. In most cases, the topics and principles are applicable across many different scenarios — from road to sea transport and production equipment to rail systems. This book may save you money, time, and effort managing and maintaining your fleet. It should also help you improve efficiency.

The channels by which money can be lost in the initial purchase of vehicles or in maintaining a fleet are many; and they are touched upon at various points throughout this text. I discuss methods that potentially could be used to identify and resolve these losses, and to review the benefit-cost considerations, prior to making decisions about how to proceed with fleet modifications or purchasing of additional vehicles. These solutions are often obvious, but not so often implemented or appreciated during difficult economic times.

Unfortunately, implementing anything from the smallest 5-minute task to the largest multi-month/year project requires an investment of both effort and time. Nothing is free and indirect costs must also be given due consideration in all aspects of your fleet management. Without such consideration, these costs can rapidly increase on an annual basis. Among the biggest expenses are those involved with the operation of vehicles. Sometimes the methods by which these expenses can be reduced can also give secondary savings from other budgets. This also has to be appreciated by senior management and taken account of when reviewing performance for different departments throughout the year.

The $ and € symbols are fully interchangeable in this text. In any specific example, the important aspect is the underlying point, so feel free to think in terms of any currency you wish. Also, some of the names, locations, and project descriptions have been changed to protect the people or organizations involved, especially because many of the examples derive from persons still serving in businesses or militaries worldwide. The details of specifications from technical military documents, such as STANAGs, have also been altered or omitted for security purposes.

This book is intended simply as a sharing of experience for the benefit of others involved in the industry. If you wish to delve more deeply into any specific area, please seek out any of the reference titles listed at the back of this book. Because of the changing nature of technology in general and vehicle technology in particular, learning is a never-ending process throughout our careers. As such, feel free to adapt and improve upon any areas discussed to better fit your particular situation.

I truly hope you will find this book beneficial.

# Part One

# Vehicle Purchasing

# Chapter 1

# General Considerations

Significant effort and work are required to ensure that a decision to purchase vehicles — and the process involved in any such purchase of new vehicles — does not lead to mistakes, excess costs, or having a vehicle in operation that fails to meet expectations. A lot of this effort is similar to that conducted during the management of any type of project. Yet, the immediate and follow-up financial implications for your business or organization can often be many times larger with respect to a vehicle purchase program. In an ideal world, when an organization needs to buy a fleet of vehicles, it would always be possible to procure a number of vehicles outright for thorough testing prior to deciding which one to finally purchase. However, in the real world of tight profit margins and limited budgets across both the private and public sectors, this extensive testing is often not viable — or, in many cases, it is simply disproportionately difficult to justify to an organization's accounting department or corporate management. As such, this book details an alternative method of vehicle selection, one that does not involve testing to destruction and the consequent necessary purchase of test vehicles.

The method of purchasing a new vehicle or fleet of vehicles, as detailed throughout this book, involves the allocation of points or marks based on the priority allotted by your organization to a number of key areas. This will ensure that your organization receives the best overall end product based on present technology and market availability. The overall process is as applicable to the construction and supply of sea-going vessels, aircraft, large equipment projects, or a fleet of trains, as it is to on- and off-road motor vehicles.

Suffice it to say that the expansion of a vehicle fleet, modification to the composition of that fleet, or the purchase of a new fleet can have far-reaching negative consequences for any organization if not carried out both carefully and correctly. The detailed process described here is for use on tender purchases of multiple vehicles or an entire fleet; as a general rule of thumb, it is not as applicable for purchasing quantities of less than 10 road vehicles. As with everything in life, however, there is an exception to this. If the purchase is going to tender or there is a high price per unit — as may be the case with some large specialized industrial vehicles, or with respect to expensive rail or sea-going vessels — then this process can still be implemented economically.

Usually, only the purchase of buildings or major production machinery will compare on the price scale with the purchase of a fleet of new vehicles. In addition, the majority of the capital cost of a major vehicle purchase is what we know as sunk or stranded costs. These are past costs that cannot be recovered, except over the entire service life of the vehicles.

The depreciation of our new vehicles cost has to be considered regarding the manufacturer's expected lifetime of the vehicles, vis-à-vis the purchase cost. A declining balance depreciation is often more accurate for a vehicle fleet purchase, as it is widely accepted that vehicles depreciate more in value during the initial years of their life. Using declining balance depreciation, the speed of depreciation is higher in the early years of an asset's life cycle as more depreciation is charged against it, and it steadily reduces to smaller amounts as the asset ages.

In order to reduce the long term impact on our annual budgets, we may decide to depreciate the entire cost of a vehicle purchase over a shorter period of time than the full vehicle life. By doing so, we can account for the capital expenditure and remove it from our annual financial statements. However, we are still not physically recovering the full capital purchase cost — the money spent on our new vehicles — until further into the vehicle life.

By using a shortened depreciable life, we are in effect offsetting the purchase costs by using other money being generated within our business. Considerations surrounding how we decide to depreciate our fleet also have longer term accounting implications for the vehicle purchase and the financial health of the company in the medium term. Therefore, we need to ensure that we have an accurate picture of the future costs involved with our potential new vehicles. It is essential to ensure, insofar as possible, that we have minimized any unforeseen costs. A company's decision to depreciate the total value of a vehicle over the first few years of its life is often made for accounting purposes: a) to show a lower capital value within the company or b) to eliminate the ongoing capital expenditure output from the company accounts and operate the vehicles as neutral assets, with only running costs to offset against.

The means by which you intend to depreciate your vehicle's value also affects how you schedule and manage your future fleet replacement program. If you depreciate your vehicles over a very short period of time, perhaps a few years, then financially they can be replaced sooner than if they still hold an ongoing depreciable book value. If you are depreciating your vehicles over their full lifetime, however—perhaps 20 years—but you decide to sell them earlier than this, then you may need to sell then for a specific price upon replacement in order to break even on their purchase value. Slower depreciation may require you to schedule vehicle replacement for later in their life when they will have been operating less efficiently and reliably for some time, with accordingly higher operating costs.

Total Life Cycle Costs have to be estimated and taken into account prior to purchase. Approximating these costs in as much detail as possible prior to a purchase will give your organization a significantly improved picture of how much effect the operation of a particular vehicle type will have upon your annual budget in the future.

From the outset of any new vehicle purchase project, up to 40% of the total available financing for the project may have to be offset for ancillary requirements. These are requirements that will allow the smoothest entry into service for the new vehicles. They may also be requirements that will facilitate the effective operation and use of the vehicles as quickly as possible after purchase. Examples include vehicle housing, spare parts, and maintenance equipment and facilities, among others. Ancillary requirements are discussed further in Chapter 5.

Travel to the vehicle manufacturer's facility is often a necessary expense during various stages of the purchase process. In addition, financial allowance must also be made for the initial purchase of logistics items such as maintenance equipment and spare parts. Training for vehicle operators and maintainers and, in many cases, appropriate vehicle housing, upon delivery of the vehicles, must also be considered. Without sufficient budgeting for these matters from the outset, the rollout of the overall project will create a negative effect upon the operation of your existing fleet infrastructure.

It is amazing to see, after decades of service, the condition of a vehicle that has been kept housed when not in use, compared to a vehicle that has been parked outside in the elements. Water, hail, snow, UV light, heat, and cold are all corrosive elements contributing to wear and damage of a vehicle body and its components. Expenditure on ancillary considerations such as appropriate vehicle covering or housing must be viewed, therefore, as a means of safeguarding for the long term — the finances being spent on the often high unit cost of any new vehicles themselves.

In my experience, many temporary flexible vehicle coverings are inappropriate in wet or damp climates. Even the supposedly breathable coverings made of various materials can often trap moisture long enough to ultimately hasten corrosion of the vehicle body and components. Such coverings in wet, damp, or humid climates should only be used for single periods of less than 24 hours, if at all. As such, the investment in solid vehicle housing — of a type appropriate to the operating climate — is a necessary cost of fleet purchase. In the case of many civilian logistics industries, this requirement may not be necessary because the vehicles are being used so intensively that their expected life is not sufficient for climate considerations to have an effect. But in the case of specialized equipment and vehicles with a higher unit value, or military equipment (especially armor), appropriate housing is absolutely critical in many climates.

Keep the following in mind when proceeding with a vehicle purchase program. Due to the high unit cost of military vehicles, specialist vehicles and equipment, and large commercial vehicles, significant savings can be realized through economies of scale. Purchasing 30, 50, 100, or more vehicles at one time allows for a single production block in the manufacturer's factory. Thus, instead of purchasing just 5 or 10 vehicles each year, we can achieve a sizable decrease in unit cost. A saving of even a few thousand dollars per vehicle is very significant when purchasing large vehicle quantities.

Of course, the viability of large bulk vehicle purchases depends upon your immediate vehicle requirements. If your requirement is simply to replace five vehicles as part of an ongoing purchase and replacement cycle, then a large purchase does not make good business sense. Also, if the vehicles are not going to be required in the near term, bulk purchasing is not the most sensible practice. If your business is not large enough to replace a similar-sized block of vehicles in the future, when these age and wear, then smaller purchase quantities — which can be replaced a handful at a time, perhaps on a yearly basis — may fit better into your accounting practices.

In the vast majority of situations, purchasing vehicles for storage, pending a requirement for them later, is neither viable nor sensible. This is because technology is currently advancing at such a rate that the savings from a bulk purchase may be outweighed by greater fuel efficiency and performance on newer vehicles that are purchased later. (These advancements are driven not only by competition among manufacturers, but also by changing legislative requirements for emissions and efficiency standards.) The threat of impending obsolescence for components is also a factor as vehicles age, whether or not the vehicles have been used much — or at all — subsequent to purchase. Furthermore, capital that is tied up in vehicles does not generate a return for a significant time if the vehicles are held in storage.

In the military, this situation can be different. The military often has unique requirements for strategic reserves of vehicles and equipment that are stored until required. In the vast majority of cases, these fleets are decommissioned vehicles and equipment that can be brought back into service. However, in some rare cases, they are reserves of new or almost-new vehicles, which if necessary are ready for operational use. Such vehicles may simply require some minor in-service modifications that their sister vehicles in use have already received. Due to the vagaries of public funds and finances, the storage of vehicles can result from purchases of new vehicles with very high operating costs, such as may be the case with Main Battle Tanks. After purchase they are not used, or minimally used for training, until required in wartime or on operational duties. In the meantime, training simulators are used to save on fuel costs and the vehicles are effectively in storage.

Vehicles are often purchased from foreign countries. Alongside possible savings from economies of scale, there can be significant savings or losses due to fluctuating international currency exchange rates over the course of a tender process. Therefore, you should lock in the purchase price at the earliest possible desirable point. A minor change in exchange rates over the course of a few months on a multimillion dollar project, can cause a difference of tens or hundreds of thousands of dollars in the final price.

Your company may have a clearly defined singular requirement when considering a vehicle purchase and such a requirement could make all other considerations irrelevant. If so, then there is little need for you to undertake a full purchasing project and vehicle evaluation. Some of the finer details of the purchasing process described in this book may therefore not be as relevant to your organization's specific circumstances. For example, your goal may be, "We require an inexpensive vehicle that can carry two persons and has the longest range possible." In this case, you would simply purchase the least expensive vehicle with the longest range and sufficient seating, regardless of any other considerations. However, this is very rarely the case, especially in specialist vehicle purchases where many conflicting requirements often conspire to complicate an already detailed process.

**A new vehicle can never be all things to all people. Compromises will need to be accepted and made.**

No product will be perfect in all areas. The vehicle you ultimately choose to purchase needs to reflect the present priorities of your organization both accurately and realistically. Insofar as possible, it also needs to account for future requirements. When we say "realistically," we refer to the current state of technology; requiring a potential vehicle to meet standards or specifications that are as yet unproven in use with other companies or organizations can ultimately prove to be a very expensive endeavor.

## Chapter 1

In most vehicle purchasing projects, the manufacturer will not be local or have a production facility within the same country or state. The experience on the next page highlights the importance of restricting yourself in your vehicle, equipment, or fleet purchase, to combinations of vehicle specifications and equipment that have already been proven — or at an absolute minimum, that have completed comprehensive and verifiable practical testing.

The first half of this book, discussing vehicle purchasing, is not about research and design; in many ways, it is about how to avoid R&D in order to maximize the savings, reliability, and cost effectiveness of a new vehicle. The ultimate aim of any vehicle or equipment purchasing process must be to achieve the most seamless introduction to service possible.

In the late 1990's a South American purchase of transport vehicles took place, which involved seeking offers for off-the-shelf vehicle designs which were already in service. NATO STANAGs were being used to benchmark the standards during the time of the purchase. A decision was made to require loading equipment to a STANAG level, which was not present at that time, on any other vehicle of the specific type worldwide. The supplier met the required equipment standard and delivered the vehicle, but the mechanical changes required in order to accommodate the weight of the modified equipment caused significant delays and changes in the fundamental design of the vehicle (shape and size). Therefore, the overall production schedule needed to be changed. Although there were some other project issues, the essential modifications deriving from this single requirement effectively resulted in delivery of an almost new, untested vehicle. This requirement and some other changes, subsequently resulted in a disproportionate number of warranty issues and spare parts delays after delivery. Luckily in this instance, the manufacturer was based in the same country and the issues were addressed in due course. However, this proximity of manufacturer to purchaser will not often be the case.

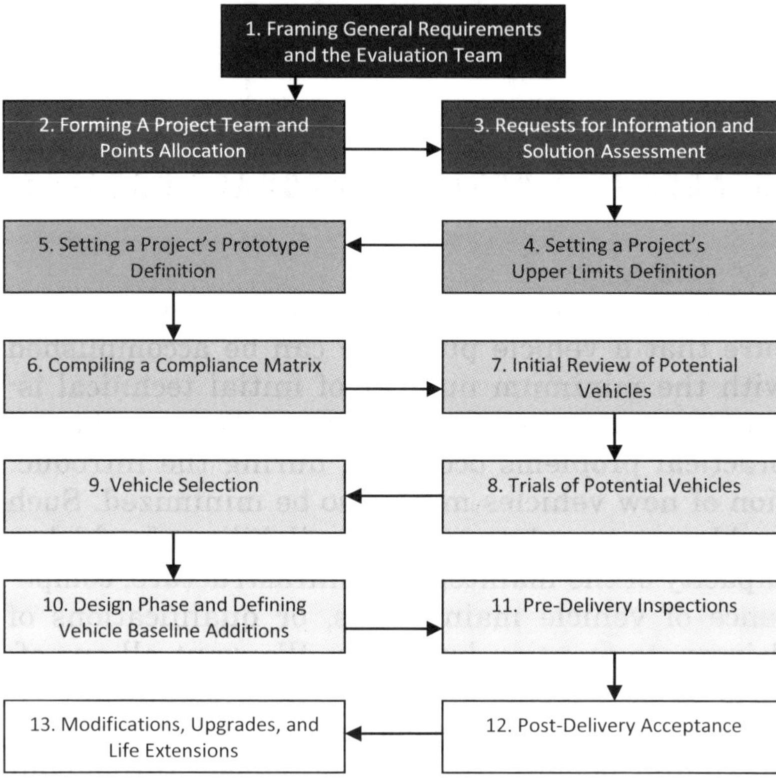


***Figure 2-1 Stages in the vehicle purchasing process.***

process and will be discussed in more detail in Chapter 4.

- **Upper-Limits: A definition of component assemblies the project team is unwilling or not-permitted to change; the maximum specifications that the project is able to accommodate.**

- **Prototype Risk Values: Numerical values set to limit the scale of engineering changes permitted on a project vehicle, to avoid venturing unintentionally or too far into Research and Design (R&D).**

- **Compliance Matrix: The minimum and target specifications which the project either desires or considers essential, and which should be achieved by any prospective supplier.**

It is imperative from this point to highlight the importance of the timeline in any vehicle procurement and purchasing project. If carried out correctly from its initial stages, the procurement process is exponentially longer with a specialist vehicle purchase. However, it can also be quite long with a standard commercial vehicle, even more so if a fleet is being purchased. For specialist or military vehicles, the entire purchasing process can often take a number of years to reach completion. It will also involve a much more thorough examination of all the vehicle type, specification, and purchasing options available.

The key point regarding purchasing timelines is that setting timelines that are too short, or driven by budget dates or organizational politics, is not conducive to maximizing financial value or to purchasing the best final product.

When we discuss timeline length, everything is relative. A purchase program for a new 4x4 fleet of jeeps or SUVs, based upon a commercially available vehicle, could conceivably be fulfilled in a period of less than a year, including all necessary tests and examinations. However, a purchase program for a fighting vehicle or military fleet can take a number of years. A program involving design and production of a vehicle from concept through prototype to final product can even take up to a decade or longer.

The length of the overall procurement process will ultimately depend upon the complexity of the final systems involved and the type of vehicle system you need to purchase. If your organization's goal is to achieve value for money and also purchase the best possible vehicle, then anything less than approximately 12 months from initially seeking proposals for new vehicles to the delivery of such vehicles, is too short for a soft skin (non-armored) commercially available vehicle purchase program. Anything less than 30 months is an excessively short timeline for a military fighting vehicle purchase program. These are minimum figures and can easily increase; *ultimately,* they depend upon the complexity of the project and the considerations that will be discussed throughout the following chapters.

## Stages in a Vehicle Purchase Project

*Since most people don't have the strategic vision and single-mindedness to pave a decade-long path, those who do can sometimes work wonders.*
*(Michael Dell)*

Whether or not a purchase is for a fighting military vehicle can dramatically affect the rollout timeline and introduce additional stages in the specification, testing, purchasing, introduction and post-delivery processes for your new fleet. The usage profile for a fighting military vehicle is significantly different in comparison to a non-fighting vehicle, specialist commercial vehicle, non-specialist commercial vehicle, or private vehicle. Any purchase project timeline has to take these differences into account from the outset. Even the best laid-out plan will inevitably meet further unexpected issues and require leeway in the final project timeline to facilitate unforeseen delays.

The following thirteen stages should be considered with any purchase program for a new vehicle. On a comprehensive and well run project, each should be completed in sequence.

### Stage 1: Framing General Requirements and the Evaluation Team

The first stage in a vehicle purchase program of significant size, or any large project for that matter, is usually carried out by senior management. This stage will often take place upon the recommendation of experts with individual specializations, working at various levels in the organization. These experts, overseen by senior management, will comprise an evaluation team. This initial stage of the process seems obvious. It simply involves the identification of a definite and clear requirement that needs to be met by the project.

Although this sounds simple, in reality it can be one of the most difficult facets of a project to tie down —especially if there are many different vested interests in the project, each with their own requirements and priorities. Such vested interests can be many and varied depending upon the nature of your organization. They may take the form of individual departments wanting a vehicle to meet requirements or technical specifications in their own area of expertise, or perhaps an accounting department wanting the vehicles to remain under a specific price. There are many other potential conflicting and contradictory requirements that can be present when purchasing new vehicles.

**Establishing General Requirements**

It is very important that at this stage in the process the framing of the initial general requirement be approached from a completely neutral perspective. Not only neutral with respect to any singular requirements from any of said vested interests, but also essentially, without any desire to simply purchase something because it is viewed as better than not purchasing anything at all. This stage in the process will ultimately decide whether or not any purchase will be pursued.

This point is very important: Purchasing something simply because the opportunity or finance exists, when the requirement for it has not been properly identified or defined, is hugely detrimental to any organization in the longer term. Finances are always going to be finite. It is essential to ensure that where money is being spent it is the best possible use for limited funding.

> *There are always more ideas for projects than there is capacity to do them. So, before starting any project, it's a good idea to confirm it should be done. Project selection is an important part of an organisation's governance.* (Newton, 2011)

All available data and information must be gathered regarding why it is necessary to purchase a vehicle in the first place. This information can take many different forms. For example, it can consist of hard statistical data from fleet analytics, outlining the present fleet running costs and potential savings from newer, more efficient vehicles. This information can also outline the potential savings from changing to higher or lower cargo capacity vehicles. Perhaps it will detail the maintenance costs on the current (older) fleet and how they have risen during the vehicle lifetime and are now inordinately high.

But most important, the framing of the general requirement for a vehicle purchase should review in detail whether a requirement actually exists. It is not unusual for projects to be initiated simply based upon a general opinion and perception that a requirement exists. Subsequently, the case does not stand up under detailed scrutiny.

Having established that a definite and clear requirement exists, and that the purchase of a new other value depending upon your project needs. A higher total number of points allows for a finer and more detailed marking scheme to be used as required.

Table 2-1 Sample Allocation of Points

- Vehicle Purchase Cost/Price **(15 – 20% of total points)**
- Mechanical Performance **(30 – 40% of total points)**
- Mobility Performance **(<35% of total points)**
- Spare Parts and After Sales Support **(>10% of total points)**
- Other Areas (ability to facilitate nation or company specific equipment, ability to integrate with existing fleet assets, role expandability, etc.) **(<25% of total points)**

Whatever the number of total points you decide to use in scoring competing vehicles — 1000, 20,000, or any other — it will always equal 100% of the allocable marks. Each vehicle will be individually marked from a total of 100%, with the final total of awarded points for each vehicle being compared at the end of the selection process. The total number of points should ultimately be allocated between the areas listed in Table 2-1, according to your own project priorities.

Suppose four vehicles are being evaluated with the aim of purchasing the single best vehicle overall. You decide to use 10,000 points as your total, so 100% = 10,000 points. You then allocate these points to the areas described in Table 2-1, according to the breakdown of percentages you decide. Again, you settle on the value of these percentages to reflect the importance of each specific area to your organization's requirements. The percentage values allotted in Table 2-2 are for example only. Table 2-2 also shows how the chosen percentages for each area directly relate to portions of the available total points.

Table 2-2 Sample Allocation of Points

| **Total Available Points = 10,000** | **= 100%** |
|---|---|
| Vehicle Purchase Cost/Price | 15% (of total available points) |
| Mechanical Performance | 40% " " |
| Mobility Performance | 25% " " |
| Spare Parts and After Sales Support | 12% " " |
| Other Areas | 8% " " |
| | 100% |
| Therefore, | |
| Vehicle Purchase Cost/Price | 1500 pts |
| Mechanical Performance | 4000 pts |
| Mobility Performance | 2500 pts |
| Spare Parts and After Sales Support | 1200 pts |
| Other Areas | 800 pts |
| | 10000 pts |

As we will discuss in Stage 8, *Trials of Potential Vehicles*, these points within the areas from Table 2-1 are broken down further into detailed nology and is simply not available for their vehicle. We need to avoid situations where the project team creates any specifications or defines their purchase requirements using limited information or, worse, within an information vacuum. They must be fully aware of the available technology and of all possible market options available when the compliance matrix is created.

**The key purpose of an RFI is to fill any gaps in the information jigsaw.**

**Working with RFIs**

In many cases, employees on a project team (or within the organization as a whole) believe they are aware of all relevant market information regarding current vehicle designs, technology, and options. RFIs can quickly show this belief to be false. They can significantly improve the initial stages of a project, at virtually no cost to your organization

In order to be clear on the importance of RFIs, let me explain the consequences of failing to carry one out at the beginning of each new purchasing process. An RFI conducted a few years beforehand, perhaps when you initially considered the purchase, is not sufficient. Not having a complete knowledge set prior to seeking offers for a vehicle purchase can easily lead us to unintentionally specify in the compliance matrix or larger purchasing/tender specification a requirement that exceeds the technology readily available in the marketplace. This disconnect can cause a variety of problems.

The obvious risk is that no vehicles will be offered that can meet all of your requirements. We will then be forced to amend the specifications — and perhaps even suffer embarrassment. Of greater concern is that to avoid saying that they are unable to supply the requested vehicle (perhaps out of concern that competitors may meet all requirements), some manufacturers may state at the outset or some other relevant stage that they expect to meet your specifications before the completion of the vehicle selection process. We must always be wary of any unproven vehicles or technology.

*New Vehicle Designs and Technology*

We are not interested in R&D. We want to keep our direct and indirect costs down by purchasing proven technology that will enter service as quickly and smoothly as possible. The only way to do this is to make purchases that have already resolved any introductory problems with other buyers. Being the first to purchase a new vehicle design or technology is not conducive to a simple rollout. Although this is less of an issue for standard commercial road vehicles, it is a very significant consideration for more unique and specialized vehicle designs. I will discuss this point further in Stages 10 and 13 of the purchasing process; however, this section contains a short explanation.

By avoiding the newest available technology, you may be concerned about missing out on cutting edge advancements before your competitors. In reality, in the area of vehicle design where the combustion engine is fundamentally the same as it was almost a century ago, significant single cost saving advancements are rare. Teething troubles and delays are often associated with new designs or technology before they operate as they should, or give the benefits which were predicted. This is fine if your business is R&D, but for the vast majority of commercial companies using specialist vehicles as tools to generate ever tighter profit margins, the risks associated with rolling out brand new designs are simply not affordable.

When being the first to use new designs, we need to reach the point where the gains outweigh the expense and effort of field testing the technology. But if the new vehicle design is suffering repeated mechanical failures and requiring repeated design modifications after purchase, then the very small potential technological gain would have been outweighed by the speed, ease, and reduced risk of purchasing a proven product.

Perhaps most important, it has been a long time in vehicle technology since one single advancement reduced costs enough in one swoop to justify being the first to test it. Being first to test cutting edge technology comes with huge risk — one that is often unjustifiable for the tiny fraction of a percentage you might save in costs if it works absolutely flawlessly first time. Most of the methods that may significantly reduce costs, such as more efficient engines, switching to alternative fuels, etc. have already been examined for reliability, or other pros and cons throughout their development and use.

If however, you are the first to try a new version of an old solution — such as a new tire thread design — a high percentage of the risk has already been borne by others who came before you. The internal structure of the tire will be a proven technology. Although the predicted gains may not materialize, the risk of repeated tire failures and lost business will be low. When trying a brand new technology or combination of technology, however, such as a completely new tire with a new internal structure and materials, you are bearing most of the risk. If a new technology or vehicle design does not work as planned, waiting for modifications and retesting could set your new vehicle project back years. It can also lead to ongoing problems throughout the vehicle life and cause significant financial implications due to lost operational hours and having to bear the costs of repairs or modifications ourselves after expiration of any warranty.

Consider the following example on the following page based on real world projects and experience; it highlights the danger of new designs.

In military R&D, in-house technology is tested, modified, and proven prior to being put onto production vehicles for field testing. But in the case of commercially contracted vehicles, the commercial imperative to make a sale can drive suppliers to sell such technology largely unproven. I have seen this happen more than once on different vehicles projects, where problems generated by purchasing an unproven design specification were still reoccurring 15 years after purchase of a new vehicle.

If you are currently involved with purchasing vehicle fleets or large specialist equipment, or will be in the future, it is essential that you learn the difference between *additions* and *modifications*. I was once involved with a project where my lack of eloquence expressing this point, contributed to a large number of technical changes to a vehicle, with extensive implications for vehicle reliability upon introduction and many thousands of hours of work over subsequent years.

So what is the difference between additions to a new vehicle and modifications to the same vehicle? I propose the following definitions:

***Addition:*** **Any alteration to the baseline vehicle that does not change the fundamental engineered properties of its mechanical vehicle components, usually in the form of add-on or ancillary equipment. Includes mountings for ancillary equipment or crew kit.**

***Modification:*** **Any alteration to the baseline vehicle that changes the fundamental engineered properties of its mechanical vehicle components. Includes removal of moving components or their replacement with alternatives.**

If your organization is approaching a project with sufficient flexibility for research and design, then by all means make as many modifications as you wish. However, if the purpose is to buy a proven vehicle, one which will enter service as quickly and smoothly as possible, the key is to avoid modifications.

## Project Additions or Modifications (Definition)

To ensure a smooth vehicle purchase, specify what you need early in the process. Then don't change the basic mechanical fundamentals. If changing the fundamentals is necessary in order to achieve the required specification, simply understand that the project is wading into R&D — and make the necessary allowances. Such allowances will be needed in key areas such as production and delivery schedules, anticipated initial reliability after delivery, Lead Time To Availability (LTTA), and price of spare parts. Additionally, each modification will require a comprehensive risk analysis and lastly, the increased risks on the project will have to be apportioned in detail within the contract between the supplier and the customer.

In many projects, opportunities will arise to apply unproven ideas and technology to your new vehicles or fleet. Even the most professional and competent staff members or project team members can become over-enthusiastic about the potential results of such ideas. The nature of such ideas can be as varied as the imaginations of the people who invent them. They can relate to alternative fuel systems, alternative power plant or drive mechanisms, moving components, reorienting components, or any other area you can think of. They may be recommended by the vehicle suppliers or from within your project team. There is also the possibility that the project team members can be so overwhelmed with the workload at particular stages during a project, they agree to pursue potentially damaging paths regarding the types or levels of technology to be pursued.

A Belgian colleague once attended presentations regarding modifications to FV106 Samson Tanks from the CVR(T) family (Figure 3-1). He related that the presentations proceeded well, but ultimately one supplier recommended the fitment of an new filtration exhaust system to the fleet. Notwithstanding that there are many positive aspects of being environmentally conscious, this particular recommendation was for the fitment of a system which at that time, was not currently in operational use with any armored military vehicle of this type worldwide. Fitting an untested and unproven system like this to the Belgian vehicles would very clearly have resulted in a situation where the customer was field testing technology in a new operating environment — and also would very likely have been involved in future modifications and R&D. Of course, the system may or may not have subsequently proven an excellent investment.

***Figure 3-1 Scorpion Combat Vehicle Reconnaissance (tracked) during an armoured vehicle shoot in Wicklow, Ireland***

Note that the technology involved in this example, and the associated recommendations, may have worked out excellently in the long term. However — and this is a big however! — the key point is that this particular system, on this particular type of vehicle, had not already been proven in operational use by other organizations. In the vast majority of circumstances, you do not want your organization to be the first operator of a new technology without very strong justification and an early acceptance of the possible failings of that technology while you "iron out the creases." In this example, the specific technology in question was not pursued at that time.

These potentially damaging paths, or vehicle modifications, can originate in many ways. They may be recommended by a supplier who wants finally to have the opportunity to say their new idea is in use with a customer. (This was likely the case with respect to the CVR(T) Light Tank example above.) These paths could also be recommended by project team members who require a modification in order to push through an agenda related to project priorities in their own area. Yes, there can be micro-agendas within any project team!

When we discuss modifications, we are not simply referring to changing or switching-out a component A for an alternative component B. For example, choosing a similar winch from a different manufacturer should not raise huge modification concerns (although there are other concerns that need to be addressed with such choices). Such a change may result in small production delays as mountings or plugs, etc., are changed on the vehicle; still, they are less serious because they do not usually cause other secondary effects or have unintended mechanical consequences for other vehicle components and assemblies. Instead, we are referring to modifications to component A that result in a new version of that part: component A-1, if you will.

When considering engineering modifications to any vehicle or large piece of machinery/equipment, the following rule of thumb can go a long way to preventing later engineering issues:

**It is a bad idea to modify a component or assembly if there is separate component that depends upon the unmodified properties of the original.**

In short, the more sub-assemblies that depend upon a component you are considering modifying, the more problems that can potentially arise and the less willing you should be to make those modifications.

In a project involving multi-purpose or multi-role vehicles, the line between additions and modifications can easily be blurred to suit the needs of specific departments within the company. This line can even be blurred to the benefit of individual departments, who are directly contributing to the selection of a particular vehicle. This blurring could also occur to the detriment of other elements involved in the vehicle selection.

At some point in the purchasing project, it will always become apparent that a specific decision has to be made with respect to which vehicle, or which specification, to purchase. At this point, when different departments are debating on behalf of their own interests, human nature can make it very difficult for a clear distinction to be defined between additions and modifications. Stage 4 of the purchasing process focuses on Upper Limits and outlines a means of controlling modifications on your project.

The purchasing team and senior management must have a very clear understanding that any modifications to a manufacturer's baseline that do not already exist on a vehicle in service somewhere in the world, fit very clearly into the concept of a prototype — and should be viewed as such. Any modifications that have a new or unique configuration, and that are simply unproven to a sufficient extent, also fit into the concept of a prototype. Single components can be proven. However, do not mistake this with a proven complete vehicle when all of

crease. For each step further away in the supply chain that the sub-assembly and minor or major component manufacturers are located from the customer, the timeline and costs grow more and more for any modified component.

The theoretical example in Figure 4-1 highlights the potential of almost exponential supply cost and time increases that a modification can inflict upon a project. It illustrates the importance of placing upper limitations on any technical project modifications. I hope it also hammers home the risks of exceeding these reasonable technological limits for your new vehicles.

Let's first consider a standard unmodified component. According to Figure 4-1, this theoretical component is sold to you for $3000, with a lead time of two months (8 weeks), containing sub-assemblies stretching back four links in the supply chain.

Now let's say you decide to modify your vehicle and that this component modification requires the suppliers at links 4 and 2 to custom manufacture parts.

Although these manufacturers are not the only ones to face cost increases as a consequence of your modification, they are the ones who will likely have the largest cost increases. Suppose their in-house production costs increase fourfold. They increase their prices further, knowing they are guaranteed the sale of these modified parts because you have stipulated this modified specification especially for your vehicles.

For the final modified component, each manufacturer in the supply chain likely has to develop

| FINANCIAL | LEAD TIME |
|---|---|
| Link 4 charge their customer (Link 3) $50 ($25 manufacturing cost = $25 profit) for their component. | 1 week |
| Link 3 conducts assembly and charges $600 (Purchase price of $50 + in-house costs of $275 = $275 profit). | 2 weeks |
| Link 2 conducts assembly and charges $1500 (Purchase price of $600 + in-house costs of $400 = $500 profit). | 3 weeks |
| Link 1 (Vehicle Supplier) fits to vehicle and charges $3000 (Purchase price of $1500 + in-house costs of $500 = $1000 profit). | 2 weeks |
| To you the customer: **$3,000** | **8 weeks** |

***Figure 4-1. Unmodified component costs and lead time.***

custom tooling or alter production procedures. This will often be the case even if they are not directly re-designing their specific sub-assemblies. Changes made to components and sub-assemblies at one stage can require modified tooling, jigs, or software code for production machinery at every subsequent stage in the supply chain. Therefore, in order to allow for modifications on sub-assemblies from their own suppliers further back in the supply chain, such production changes will likely be necessary. All of these changes take time, increasing costs for the suppliers and, ultimately, the price of the final parts. At the very least, each link in the supply chain has to re-evaluate their own assembly procedures to ensure quality control on the altered parts and sub-assemblies they are receiving.

The suppliers at links 4 and 2 in Figure 4-1 are now custom manufacturing parts. In Figure 4-2, we can see that the suppliers at links 3 and 1 (working back from the final customer) incur double their in-house production costs due to the changes to the parts they are being supplied.

| FINANCIAL | LEAD TIME |
|---|---|
| Link 4 now charge their customer (Link 3) $200 ($100 manufacturing cost = $100 profit) for their component. | 4 weeks |
| Link 3 now conducts assembly and charges $1500 (Purchase price of $200 + in-house costs of $550 = $750 profit). | 4 weeks |
| Link 2 now conducts assembly and charges $6200 (Purchase price of $1500 + in-house costs of $1600 = $3100 profit). | 12 weeks |
| Link 1 (Vehicle Supplier) now fits to vehicle and charges $12400 (Purchase price of $6200 + in-house costs of $1000 = $5200 profit). | 4 weeks |
| To you the customer: **$12,400** | 24 weeks |

*Figure 4-2. Modified component costs and lead time.*

In the worst case scenario, every link in our supply chain has to modify their assemblies, adjust or replace equipment to accept such modified sub-assemblies of altered shape or design from their own suppliers, or redraft and reissue modified assembly drawings and procedures. The costs for all of these changes and the time it takes to implement them at every level are passed on to your modified components.

This example is extreme, but serves to illustrate the point. You can see the increase in component price from $3000 in Figure 4-1 to $12,400 in Figure 4-2, with an accompanying increase in lead time from 8 weeks to 24 weeks. This modified component could be anything you change the specification of on your vehicle — from a winch, to the steering assembly, braking system, axle, to an air conditioning system, etc.

Another important issue in modern supply systems for vehicle manufacturers further complicates the matter of component modifications. The example above does not take account that the modern supply chain for a major assembly is likely to be shaped more like a branched tree than a linear chain. The vehicle manufacturer is quite probably conducting assembly on more than one sub-assembly in order to create your modified component. Each of these suppliers at the very end of the supply chain may also have a number of their own raw material suppliers and so on.

If you have modified a component or assembly on your vehicle; it may now contain custom made sub-assemblies manufactured a number of steps earlier in the supply chain. This will hugely increase your final in-service component Lead Time To Availability (LTTA) and subsequent unit replacement costs for the entire life of the vehicles, as shown in Figure 4-2. The theoretical increased lead time outlined above demonstrates delays only in the supply of components. The actual initial delivery times for the new vehicles are further increased by the testing and certification processes necessary for new designs or modified components. This is especially the case if key safety areas are being modified — areas such as brakes or steering, which will require an extensive and legally binding testing and certification process.

In military vehicles — indeed, in any vehicles — Commercial Off The Shelf (COTS) components are the key to low cost, sustainable maintainability. Military Off The Shelf (MOTS) components often can't compare to the advantages of purchasing a vehicle that maximizes the use of high quality COTS components. However, either is much better than custom made or uniquely modified components.

Further increasing the costs of any components that are specific to a modified individual vehicle design are the raw material purchase programs of large vehicle manufacturers. Modern manufacturers often purchase their raw materials in the form of bulk metal contracts years in advance, based upon complex projections of commodity prices. They may then hedge the value of these materials against purchases of futures contracts in the international stock markets. Digging into the detail of these processes is beyond the scope of this book. However, I need to highlight that any unique requirements that a customer imposes on a vehicle design, and which may interfere with such projections, can increase in cost during the initial months or years of modification research, before the vehicle has even entered into service.

Worse, the modified component costs can escalate within a few years, when you are seeking to purchase quantities of essential but unique spares after the delivery of your company's unique vehicles. Unless a fleet consists of hundreds or more vehicles — where these risks are offset by the capacity for bulk purchase of spare parts in the future — cost escalation is yet another reason to avoid component modifications.

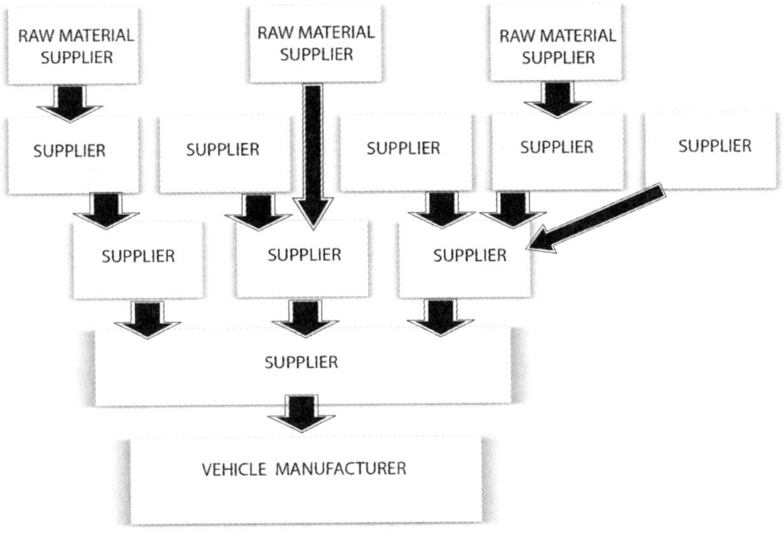


*Figure 4-3. Sample single branch of a vehicle component production supply chain.*

Figure 4-3 demonstrates the nature of modern supply chains. Such diagrams usually appear more like branched trees than a linear chain of supply. This example shows only one component supplier whereas a modern vehicle manufacturer may have hundreds of direct suppliers, each of whom can have numerous sub-component suppliers.

## Stage 4: Setting a Project's Upper Limits

What is an Upper Limit for a project? I've coined this term to describe the maximum technological boundaries beyond which a project team cannot venture with respect to their new vehicle or equipment. These limits should be specified early during any vehicle or large equipment purchasing project. They ensure that the project does not overstretch regarding technology, expense, or time.

In my experience, there is often very little appreciation for the huge benefits of setting maximum boundaries for any project prior to its embarkation. The focus is generally placed upon the project's minimum requirements. Upper limits are not given much thought until budget overruns or increased lead times force their consideration. There is also little appreciation for the importance of such limits regarding the technical issues they can prevent on new vehicles.

Unfortunately, in any organization, every project will be subject to the ebb and flow of human nature. Sometimes members of a project team, especially on the non-technical side, can fail to appreciate the scale of changes being proposed to a vehicle or equipment project's baseline. This failure is not so much a failure by the team members themselves, but can be attributed to flaws in human performance factors, *vigilance decrement* (cognitive fatigue), or the limits of the conscious workspace. A purchasing project should always have maximum limits in place from the outset.

*"In the case of…errors…the defences are the system safeguards and barriers that detect and recover errors before they can have a damaging result." (Reason and Hobbs, 2003)*

Your upper limits for a vehicle or equipment technical project can generally be defined as:

**The maximum technical specifications of assemblies or features without any customer modifications**

Upper limits are a more basic tool than your compliance matrix (which will be discussed in Stage 6) and serve a different purpose. They provide a project team with very clear technical boundaries. These boundaries can be referenced during the vehicle selection process and the project team must work to keep from exceeding them.

**The fundamental difference between upper limits and the compliance matrix is that upper limits focus on maximum specifications whereas the compliance matrix focuses on minimum specifications.**

In extreme cases, and if absolutely necessary, changes can be made to the upper limits during a project. However, changing your fundamental upper limits should require serious discussion, examination and review by the project team. It is not a decision to take lightly due to the potential for serious secondary effects that will affect the project. If you need to redefine your project's upper limits, alarm bells should activate within the project team because either something has gone wrong with the original tender/purchasing specification or an important detail was missed when creating the upper limits. To summarize:

**A project's upper limits define specifications which should not be changed during the project from those initially offered by the manufacturer.**

**The project's upper limits should ensure that the project team do not exceed reasonable engineering changes, either after a final vehicle selection is made, or in order to facilitate purchasing a specific vehicle.**

If, contrary to the upper limits in place for your project, a decision is made to modify standard vehicle components and increase specifications, there can be significant financial implications for your project, and hence for your company. There are two means by which you, as the customer, can avoid or minimize this initial financial burden of upper limit's modifications.

The first method is that the vehicle supplier agrees to bear the costs. This may occur if a modification is required due to an engineering design issue that is identified during the trials, or simply as a concession to help win the sale.

The second method of reducing these costs is by being in the unlikely situation where you can change to alternative proven components that are 100% compatible with components and fittings already in place on the vehicle, preferably COTS components. This opportunity is very rarely possible.

Regardless of how you avoid these initial modification costs, if the resulting changes are unique to your specific vehicle, there are long-term costs in the form of increased spares prices and increased spares LTTA; these costs will not be borne by the manufacturer. Also the time burden to a project of any modifications can also not be avoided without seriously compromising the testing and quality standards on the new units. With the onward progression of corporate manslaughter, gross negligence manslaughter, and negligent homicide legislation in many countries, such testing and certification requirements — especially for key safety-related items such as brakes or steering — are ignored at your peril. For example,

# Chapter 4

*An organisation...is guilty of an offence if the way in which its activities are managed or organised (by its senior management) .... causes a persons death; and amounts to a gross breach of a relevant duty of care owed by the organisation to the deceased." (www.cps.gov.uk, 2014)*

At the outset of your project, you will have to define your upper limits to suit your project's specific goals. Essentially, you list the major components, specifications, or items that the project team wants to avoid changing on the final product. The reasons for avoid major modifications can include keeping costs down, maintaining the desired timeline, preventing unproven engineering modifications, or combinations of the above. This definition is not specific to each individual vehicle; instead, it is a list of limitations that will be imposed on any chosen vehicle, regardless of the vehicles that are being evaluated or will be eventually chosen. After a final vehicle type is chosen, the list can still be amended to place limits on any changes to that vehicle's unique features.

For example, if you want a power source that has been proven in operational use, you may limit the fuel choice to a simple diesel or petrol engine, and not permit alternative fuels or hybrids if such fuels have not previously been used on your specific vehicle type or in your specific operational circumstances. In this case, you are defining an upper technological limit to prevent a situation where you end up testing a relatively new technology. Of course, if alternative fuels or hybrid technology have been tested and proven in use on your specific vehicle type *prior* to your purchase, this example does not apply.

The following list indicates, from the most practical method to the least practical method, potential ways to define a vehicle purchasing project's upper limits.

1. Your upper limits are based upon the specification for all competing vehicles, as seen on the day they are supplied for customer trials or testing. This is the preferential method for defining your upper limits and also the preferred time to do so. Sufficient time must be allotted at the beginning of the trials to create this specification. This time could be a number of hours or days, depending upon the complexity of the vehicles or equipment being tested. Using this method, you can accurately reflect in your upper limits the exact specification of vehicle components with respect to the vehicles or equipment you will be evaluating.

2. Ask competing manufacturers to submit a manufacturer's vehicle baseline definition at the earliest stages of a purchasing competition; use this information as a foundation for the project's upper limits. Although you will not get to physically view and confirm the received specification prior to establishing your upper-limits definition, it will give you a basis from which to work. You can then confirm the submitted baseline later through open source research, or in the testing process, as applicable. It is important to make sure that the baseline definition you receive — and de-

fine your upper limits using — lists only technological specifications that are already tested and proven in use.

3. Define the project's upper limits using general boundaries of specifications that you know any final vehicle will have. The disadvantage to this method is that your upper-limits definition will not account for the unique features on any of your vehicles. It may not accurately account for the standard specifications of an individual vehicle. As a consequence, your upper limits may have to be changed during the process, perhaps more than once — which is contrary to the principles of having an upper-limits definition.

How your organization sets its interpretation of its upper limits can be adapted, as required, so that it best suits the needs of an individual project. Generally, Method 1 or 2 above is the most straightforward to implement, with 1 being preferred. This is because you have the exact specifications in front of you to use when defining your upper limits.

Regardless of which method is used to define your upper-limits for the project, it is essential that this definition is in place prior to beginning the individual vehicle selection. The upper limits are basic nuts and bolts of the vehicle or equipment specifications. They are the standard elements in your final vehicle, those most likely to exist in any chosen vehicle. The final vehicle will almost never exceed the upper-limits definition, regardless of dif-

In the mid 2000's an Irish construction company was conducting testing and acceptance of a new telehandler with an upgraded load capacity. The original lift was installed in accordance with all of the manufacturer's specifications; it should have operated without issue. Despite this, the system was suffering catastrophic failures during use above a 3300kg load. These failures were occurring due to complete disintegration of a connection between a new hydraulic ram and the chassis.

After much intense communication with the vehicle manufacturer and the third-party suppliers further back in the supply chain, a redesigned joint between the ram and the chassis appeared to be the solution. The new design successfully completed extensive laboratory testing in an engineering lab. It was then duly recommended for a full rollout and fleet-wide fitment by the telehandler manufacturer.

Concerned that leaking and some cracking issues were not the root cause of the failure, and despite significant timeline pressure to provide a resolution to the problem, the fleet-wide fitment was rejected by the construction company team liaising with the manufacturer. The customer insisted upon the new design being fitted to a test vehicle and undergoing further practical testing. Within 6hrs of actual use, the new design also suffered a catastrophic failure due to the vibration frequency and induced loads on the chassis bracket, which were not present on the test bench.

ferences between the final chosen manufacturers or individual vehicle suppliers.

Again, components defined in a project's upper limits are the fundamentals. They will always be present; the project team or company will not risk changing them during the project. These may include elements such as the engine, steering, brakes, fitted with wheels or tracks (which do you require on your final vehicle?), ballistic protection (what STANAG level is required?), mine protection (what STANAG level is required?), etc. Once you have specified these technological limitations, do not risk purchasing something that exceeds them during the course of vehicle selection *unless you are 100% certain* that the higher specification is thoroughly tested and proven — tested and proven in actual use, not just on a test bed because, as the previous example shows, the two can be very different. A further redesign ultimately solved the component problem in this example. However, as the example highlights, if you are going to make a modification to a moving vehicle component, neither academic calculations nor laboratory testing will be sufficient to confirm the reliability of the modifications in actual use. To prevent potential failures occurring after delivery of modified components or assemblies, it is essential to have the manufacturer carry out comprehensive field testing of the new design. Modified components *must* be tested in actual use or through *realistic* field testing.

| RECOVERY VEHICLE PROJECT UPPER LIMITS DEFINITION | | | |
|---|---|---|---|
| Serial No. | Major Assembly | Limitation No.1 | Limitation No.2 |
| 001 | Engine | Combustion Engine — Diesel | Diesel engine, Turbocharged, all as per original spec. |
| 002 | Brakes | Manufacturer's Standard | |
| 003 | Steering | Manufacturer's Standard | |
| 004 | Hoist | Permanently Fixed to Vehicle | Max 12 ton |
| 005 | Wheeled | 6 Wheel | Max 6 x 6 |
| 006 | Armor Protection | Grenade and Mine Blast Level 2X<br><br>Kinetic Level 3X | |
| 007 | Etc. | | |

***Figure 4-4. Sample upper limits definition.***

## Creating Your Upper Limits

We have outlined in detail why a definition of upper limits is required, but how do we actually write one up?

Figure 4-4 shows an example of a project's upper-limits definition. It appears short because the definition covers a limited number of key basics for your vehicle or equipment. These basics are the technical specifications *in key areas* which, like the risk values discussed in the next stage, introduce significant engineering risk if changed. To exceed these upper limits could cause significant delays or future costs. This sample definition is for a theoretical purchase of a recovery vehicle.

In this example, no minor assemblies are listed. For example, no wing mirrors are listed because changing the type or specification of a wing mirror will not have significant implications for the vehicle build. Similarly, no seat styles are listed because changing the type of seats being fitted will not significantly change the vehicle build (excluding blast protective seats). You apply upper limits only to those major components or assemblies that could have a significant impact on the production schedule or vehicle design if they are modified.

If necessary, the creation of the upper-limits definition can wait until after the compilation of the compliance matrix (see Stage 6). You might wait in order to ensure that the final definition sets boundaries that meet the overall requirements from the compliance matrix. However, as mentioned previously, do not wait any longer than the first day of any practical vehicle trials or testing that takes place during the purchasing process. Any later and you risk that the definition of the upper limits will be influenced by the developing trials' progress.

At the start of this chapter, I mentioned risk values for the first time. These values, which relate directly to your definition of a prototype for your specific project, will be explained in Stage.5.

However, suffice to note the following difference at this juncture regarding the relationship between your upper-limits definition and your prototype definition:

> **All of your Upper Limits should be assigned a risk value in your prototype definition.**

> **All items in your prototype definition do not require inclusion as an Upper Limit.**

The following discussion explains why each of the serials from Figure 4-4 were included in the upper-limits definition. Remember: these limits and their explanations are examples; you will need to decide what to include for your unique individual project.

*Serial No. 001*

On almost all projects, the general goal with regards to engine specification is to have sufficient power, using the simplest and most reliable technology. Too much excess power or torque in many cases will achieve nothing other than increased running costs. In this case, our RFI has also indicated that diesel engines are currently the standard for this type of vehicle.

Basic Limitation No. 1 in your upper-limits definition specifies that your vehicle should be fitted with a diesel combustion engine. This limitation is intended to prevent the project team from becoming wooed by proposals from potential suppliers to provide any new or unproven alternative engine for the vehicle. Perhaps your RFI suggested that alternative engines for this vehicle may be largely untested or simply not in widespread use worldwide. In this case, your compliance matrix will also specify a diesel combustion engine as an *"Essential Requirement."*

The remark *"turbocharged as per original spec"* in Limitation No. 2 ensures that the project team does not approve major engineering modifications in order to alleviate engine problems such as a lack of power. If a vehicle is lacking power, and the engine is in the upper-limits definition as something that cannot be altered from an already proven system, then look at alternative vehicles or investigate other solutions, such as potential modifications to non-upper-limits assemblies or components. Such modifications could resolve the problem in ways such as reducing weight; otherwise, you must review the project in its entirety.

In this example, if your proposed vehicle initially does not use a turbocharger at all, having one specified in your upper limits would not present an issue; remember that upper limits are *maximum* specifications. If your compliance matrix requirements are still achieved, then using lower specifications or less technology is always acceptable. The vehicle specifications should always meet or exceed your compliance matrix and be equal to or less than your upper limits.

**A vehicle which performs to the required standard using lower specifications does not violate the upper limits. It may, however, violate the compliance matrix, which is a separate issue.**

*Serial No. 002*

A key element in deciding what assemblies or components to include in your upper limits is an appreciation of how critical any changes to these will be. The knowledge and experience of technical personnel can, of course, be useful if required.

In this case, braking is included. Changing the braking system on a vehicle can often require extensive testing and certification to meet state, national, or international legislative standards. Meeting these standards takes time and will often push out a project schedule. If this time is not available, these changes need to be avoided.

The braking system on any vehicle is a fundamental safety aspect. It may seem obvious to readers, but clearly any proposed vehicles or modified vehicles that are put forward for purchase by your organization, without a fully certified and proven braking system, should be dismissed.

When a manufacturer proposes a vehicle to your company for purchase, there are very few reasons to want to change the manufacturer's standard braking system. One of those rare instances when this might be required is when the vehicle weight increases significantly. A significant weight increase may require braking changes to ensure the vehicle continues to meet legislative standards. However, the process of implementing physical braking modifications to a vehicle, and then also passing the necessary testing, is both time consuming and expensive — with no guarantee of passing the testing! Therefore, any significant vehicle weight increase should be avoided in order to avoid the requisite braking modifications that it will cause.

If the Upper-Limit's are specified correctly and enforced, then the net weight of any vehicle chosen should not increase more than a minor amount. If a project team fails to enforce the upper limits, then the effect of any modifications and the subsequent number of secondary modifications involving systems such as brakes can multiply fast.

*Serial No. 003*

As per braking above.

*Serial No. 004*

This example is based on the theoretical purchase of a recovery vehicle. Therefore, this vehicle will be purchased with a lifting hoist or crane (Figure 4-5).

The minimum size of hoist required will be specified in the compliance matrix. However, as explained earlier, the maximum size of hoist acceptable is specified here in the upper limits specification. In this instance, a 12-ton hoist is the limitation imposed. You will likely have determined that above 12 tons, the increased capacity will cause an increase in vehicle weight to an extent that may begin to require other modifications to the vehicle — hence, the limitation. This determination is based on your experience; the class, type, and size of vehi-

*Figure 4-5. An Italian Mercedes 30 metric tonne MHC (Mobile Heavy Crane) in United Nations livery. Although the gross weight of this vehicle is 38 tonnes, vehicles fitted with lifting equipment are often referred to by the capacity of their equipment.*

cle you are purchasing; and on what the RFI has shown is the standard sizes of hoist fitted to this class, type, and size of vehicle.

Determining such limits requires knowledge of modern market availability and familiarity with the current industry standards for your vehicle or equipment type and specifications. This information will have been gleaned from your RFI.

An example of other secondary modifications related to fitting a larger crane or lifting hoist to a vehicle may include wheel and axle capacity limitations. Perhaps the crane or lifting hoist mountings will also affect vehicle weight. Any additional weight can cause other significant modifications, as per serial No 002 and 003 above. Consequently, the amount of technical and time management issues with the project could potentially snowball if some upper limit is not placed on crane and hoist size.

If, however, a manufacturer proposes a vehicle with a 14-ton hoist as standard, and it is already fitted to the proposed vehicle type and already proven in service, then, by all means, change this upper limit specification. But change the specification upwards from 12 ton to 14 ton, *only if* any vehicle modifications that are required to facilitate the change are already in service somewhere and have a proven reliability record.

*Serial No. 005*

This limitation may seem unnecessary, but again the upper limits are the key fundamentals of the type of vehicle you hope to purchase. Deciding during a project that the team wishes to purchase a specific vehicle but change it from wheeled to tracked, or vice versa, is probably one of the most extreme and clear examples of why project modifications are usually a bad idea.

To the experienced reader, this scenario is very unlikely. Still, even the most non-technical reader can realize that significant work is involved in a modification from wheels to tracks. This would also involve fundamental hull, chassis, and drive train redesign — significantly changing the vehicle to the extent where the end product would have little resemblance to the wheeled vehicle chosen initially during trials.

Again, the project team's permission to make changes to any new vehicle should be restricted by the upper limits definition. I am continually surprised at how people can swoon in the presence of ideas that may seem nice on paper or in a presentation, but are ultimately being proposed by a company who wants only to sell a particular vehicle or product. Remember: no matter how good the relationship, the manufacturer's and supplier's key goal is always to sell their products.

*Serial No. 006*

In the military, this limitation is the big one. It is also acutely relevant in the security industry. Is it worth changing every mechanical component on a proposed vehicle in order to move one step up in the level of protection? Because armor protection is simply heavy and has a significant effect on a vehicle's net weight, this question is one that your project team may have to answer. Okay, this concern is unlikely to be raised in most civilian fields; however, it does serve to highlight the weight consideration to all readers.

If you are unwilling to significantly push out the project timelines and introduce a now unproven vehicle, then your answer is "No, it is not worth it". Such increases in armor will likely involve extensive changes to different systems (e.g., axles, wheels, brakes, steering, and possibly engine and transmission) and the overall solution. Furthermore, ongoing changes to vehicle components are likely after delivery. Again, if your answer is "No," dismiss the proposed vehicle and review the costs of purchasing an alternative proven vehicle that meets your protection requirement, without having to implement significant baseline modifications.

To the uninitiated reader, the answer above may seem incorrect — protection of life should always be paramount. Although it is outside the focus of this book, realize that in the military, protection of life involves not only a vehicle's inherent protection level, but also its mobility and firepower, the other two pillars of the protection triangle. Thus, sacrificing vehicle reliability — a significant reduction of the mobility pillar — for the benefit of inherent protection is no better than doing the opposite.

Whether it is armor protection causing a weight increase or any other modification, such as a significantly larger hoist, a balance must be struck. This balance is between the benefits you expect to gain from your modifications, the secondary or tertiary modifications that will be caused by any additional vehicle weight, and the reduction in reliability that will be caused by unproven technology or systems during a vehicle's introductory years.

Remember that the upper limits define the maximum specifications. The alternative vehicles do not have to meet these limits as the project's minimum protection requirements are defined in the Compliance Matrix discussed in Stage 6.

You should not seek to modify a selected vehicle to exceed the readily available technology in the marketplace because this will not yet be tested or proven.

**If a manufacturer states they can achieve a higher specification than your upper limit, or higher than what you believe is proven in use with other customers to date, but it will require modifications to the vehicle, then do not accept it unless the manufacturer can provide evidence the modification is already in service and well proven.**

You must not be blinded by ever-increasing specifications. Stick with the levels and limits the project team imposed a number of weeks, months, or years ago at the outset. These limitations were defined for a reason.

If the answer is "Yes" regarding whether the organization can push out the project timelines and introduce unproven modifications, then no problem. Be honest and admit that the project is now carrying out R&D and be prepared to push out the timeline to a realistic level in order to facilitate these changes. The organization should also be prepared to change the approach to the project accordingly, in order to account for the likely increase in testing and proving requirements subsequent to major design changes.

This example demonstrates that the upper-limits definition is a short table that provides the most fundamental controls and limitations for only the most major aspects and assemblies of a proposed vehicle purchase. Do not make changes to items included in your upper limits without very serious consideration!

## Chapter 4

# Stage 5: Setting a Project's Prototype Definition

### The Prototype Question

When does a modified project become a prototype?

> **Prototype:** *"An original type, form, or instance serving as a basis or standard for later stages." (thefreedictionary.com)*

Note the relationship between your upper-limits definition and your prototype definition:

**All of your upper limits should be assigned a risk value in your prototype definition.**

**All items in your prototype definition do not require inclusion as an upper limit.**

The word *prototype* can foster many different reactions depending upon the situation. In an R&D department it is commonplace, whereas in a vehicle purchase it is often the antithesis to the guiding principles of simplicity and reliability we hope to achieve.

Prototype does not have to be a dirty word, even for small organizations. There are ways by which organizations can benefit hugely, and at minimized cost, from making vehicle changes that in a literal sense are defined as prototypes. Some baseline modifications or changes to upper limits that have not previously been applied by other customers, can represent great improvements or value for money. However, early acceptance is required that every modification brings increased risk and increased financial costs to a project. They also bring extended time requirements to take the project to completion. These risks and requirements may not be immediately apparent.

There is a significant difference between dipping your toe into the lake of modifications and diving in head first! It is much too easy for a project team, and also the suppliers working with them, to begin with one or two small modifications, which are added to and extended continuously until we move towards what can only be considered a new vehicle. The specialist members of any project team have a significant responsibility to highlight when changes begin to set off silent alarm bells. Likewise, a project team leader must heed any warnings received.

Your project team needs to define clearly in advance the threshold at which the vehicle being purchased stops being the vehicle the team originally intended to purchase and thus effectively becomes a prototype. An easy, arbitrary answer could be after 50% change to the original vehicle baseline. But deciding the actual percentage will be dependant upon the nature of the vehicle modifications, the secondary or tertiary effects of any modifications, and the level of percentage changes your individual project is able to bear when all factors are considered, including increased project timelines and post-delivery mechanical problems with any new technology.

I contest that 50% change to a vehicle baseline as given above could be an extreme amount of alterations. The following questions highlight the difficulties with defining a clear line after which you are in prototype territory. They also highlight the type of thought process you must go through to reach a prototype definition.

- What is the nature of the changes which will be clearly defined as representing this percentage (50% in our example)?
- Will this percentage not be reached until every component is changed except the chassis? Or if only some components can be modified, at what point do you reach 50%?
- Can a 50% change to the baseline be exceeded immediately by any changes to the braking system, because your company weighs the importance of this system so much that it is unwilling to allow changes to it? Or perhaps a 50% change will comprise both brakes and steering system modifications?
- Is your higher organization management willing to accept the proposed percentage changes from the original vehicle baseline in your particular project?
- Will the timeline to vehicle introduction realistically allow the scale of modifications you're proposing?

- Is the end result of all modifications likely to justify the inherent increased risk to the project?
- Who will bear the costs associated with any vehicle design changes?
- Who will bear the costs associated with re-modification of failed changes?

These are all questions that need to be answered well in advance of them ever directly arising. If the project is within a large organization, is there an overarching guidance document on vehicle purchases? Does this document give concrete policy guidelines on these types of questions and the extent of modifications a project team is permitted to make on mechanical purchase projects?

**A vehicle purchasing policy guidance document removes the individual project and its team members from the equation. A technical guidance document for major purchasing projects should exist in all large organizations.**

A straightforward question, sometimes asked by non-technical persons who want to implement a modification to a manufacturer's baseline is *"Will this change work?"* or *"Can this technical problem be fixed?"*

The fact that such a question is being asked already indicates that there has been a questionable selection of a particular vehicle or item of equipment for purchase. Ironically, the technical expert's safe answer to this type of question will almost always be "yes" as explained by the alternative answer below. I suggest that when the reader hears the answer "yes" to such a question, it should be substituted mentally with the following two sentences.

**Almost any technical modification that is given enough time and money can be made to work — eventually. It may work after a lot of expense, false starts; and many years; it may not work well or to your satisfaction; but it will operate, it can be said to be fixed, it can be said to work.**

From the statement above, it is obvious that two key issues which require clear definitions and boundaries prior to your organization embarking on a major vehicle purchasing program are as follows:

- The first issue which requires a prior boundary for the project and which ties into the issue below was already discussed in Stage 4. This is where your organization is willing to draw the line, between acceptable and fundamentally unacceptable modifications on the project. The project's upper limits are the "black and white" line between these acceptable and unacceptable modifications. Be certain they are defined carefully.

- The second issue is the purchasing organization's or customer's definition of a prototype. A project team can potentially make tens or hundreds of changes to a vehicle without exceeding the upper limits. Such changes also need to be controlled and limited. Your project team may be briefed that they will not purchase a prototype of any vehicle, or such a decision may be reached by the team themselves. If a no-prototype limitation is placed upon the project, it is fundamental that the team approving changes to a vehicle baseline knows what is, or is not, a prototype. An easy definition of a vehicle that is *not* a prototype is one already in service with other customers. However, this definition will prove useful only until modifications begin to be made during the purchasing process. The team needs to know the point at which the acceptable proven vehicle undergoing reasonable modifications, becomes an unacceptable, unproven prototype vehicle *with too many modifications*. This point is determined by your prototype definition.

A project team can define a prototype by weighting components according to an evaluation of how much disruption they may cause to the proven vehicle design if they are modified. A good rule of thumb is that the more sub-components you have dependent upon a named assembly, the more you should avoid changing this assembly. These changes carry risk not only to the assembly being modified, but also to the ancillary assemblies and components. Thus, there is an inherent increased level of risk. Also, if a component is directly related to legal vehicle certifications, then there is increased risk due to the possibility of failing any recertification testing.

In order to define a limitation for the allowable scale of changes to a vehicle during a purchase project, we can draw up a table of maximum risk values using the following method.

First, an arbitrary number should be decided for this project's total level of unacceptable risk. Depending upon the complexity of the vehicle being purchased, this value should be large enough (500, 1000, 10,000, etc.) to give a project the flexibility required in differentiating levels of risk between components.

Once the total value is decided, allot percentages to different vehicle components reflecting the amount of risk that modifying them introduces to the project. For example, if the total level of unacceptable risk on our project is 10,000 points, allot 50% or 5000 points to the braking system and 2% or 200 points to the wing mirrors. As the project progresses and modifications are required to the vehicle, the values allotted to any component being modified should be added together. Once this sum reaches the total for the unacceptable risk value, in this case 10,000 points, any more changes make the risk to the overall project unacceptable. The vehicle then becomes a prototype according to your definition. Of course, defining your risk values requires a team member who understands the likely technical implications of changes to individual components or assemblies.

The project team can rigidly freeze the vehicle definition once this prototype risk value limit is reached. But even if the risk value is being used only to highlight a growing level of vehicle changes in the project, it is still advisable to follow this prototype risk value system on the project. Following the system will allow the project team to track the scale of overall changes being implemented on a vehicle. These changes will often occur over the course of months or years during the design phase of a large vehicle purchasing project; there can be an unintentional failure on the part of project teams to track them holistically.

The following example demonstrates the principles of using a prototype definition and unacceptable risk values as described.

Our unacceptable risk value in this example is 500 (Figure 4-6). This value is split between different components. Here, the lighting system is given

| PROJECT PROTOTYPE MAX RISK VALUES | 500 |
|---|---|
| Assembly | Risk Value |
| Engine | 501 |
| Transmission | 350 |
| Armour | 501 |
| Brakes | 350 |
| Steering | 350 |
| Lighting (External) | 50 |
| Axles | 250 |
| Wheels | 100 |
| Cooling System | 501 |
| Suspension | 200 |
| Hoist | 150 |
| Weapon Mountings | 150 |
| Etc... | |

***Figure 4-6. Sample Maximum Risk Value Table.***

a risk value of 50, the wheels are given a risk value of 100, etc. Initially, almost any individual components on the vehicle can be changed or modified. Each time a component is modified, however, its risk value is added to the risk value of any other modified components. Thus, we are keeping a running total of the overall level of risk on the project.

As long as the total value of all added items does not exceed our total project risk value of 500, we can continue modifying different assemblies and components.

In Figure 4-6, we have specified the engine with a value of 501. This value immediately exceeds our total project risk value of 500 because we do not want to accept the risks involved in making engine changes on any vehicle we choose to purchase. The reason may be simply due to timeline considerations, or it may be due to the significant secondary and tertiary effects likely to be caused to other vehicle systems by an engine modification or change. The same applies to the cooling system and the armor. The armor has a risk value of 501 because an increase in vehicle armor will likely cause weight increases and consequential modification requirements to a significant number of vehicle components.

The brake and steering risk values are set high, but they do not exceed our maximum project value of 500 (i.e., they are not at 501), because the project is willing to risk alterations to these. We will accept possible changes to these systems, with an awareness of the increased risk involved with testing and re-certification of the vehicle. However, because of the time involved in changes requiring such certifications and the increased risk involved due to the possibility of failing such re-certification, the project cannot also bear the weight of too many other modifications — the timeline may slip, or perhaps later re-modification will be required to the altered brakes or steering. Hence, these are given risk values of 350 which will allow other changes to the vehicles but significantly limit the overall number of changes allowed in addition to them.

**If you make any alteration to a component listed in your unacceptable risk table, no matter how minor, the risk to the successful completion of the project increases. Hence, your running total of risk values increases. When you reach your total unacceptable risk value (prototype definition value), no more changes are permitted. Regardless of risk values, if any changes are made, they cannot exceed your upper limit for that specific component.**

Again, it is important to specify this definition and the various component risk values at the outset of the project. They are intended to limit the scale of engineering involved in the project and to keep the project team within a clear set of boundaries when the tough decisions have to be made. This restriction is also key to maintaining the project timeline. The list of assemblies and components that are allotted a prototype risk value should extend to include all major assemblies for your proposed vehicle type.

## Stage 6: Compiling a Compliance Matrix

The Compliance Matrix is used to specify to potential suppliers your minimum and fundamental requirements for a project. It is created at the beginning of your purchasing project and effectively lists exactly what it is that you want to buy.

The compliance matrix should specify the characteristics of any potential vehicle which you, the customer, would find desirable and essential. The difference between these two types of characteristics is defined here:

**Desirable characteristics: Those characteristics that you would like to see on a vehicle, but are not essential.**

**Essential characteristics: The fundamental requirements that must be met, or a supplier's vehicle will not be chosen for purchase.**

Military organizations can often tend to over-spec their vehicles, but this trait is also often seen in commercial fleets, for example regarding the engine capacity of a new vehicle. The trend to seek higher specifications than necessary on a potential new vehicle is ironically only challenged by the equally frustrating capacity of purchasers to under-spec in other areas.

Generally, the trend to over-spec when deciding vehicle specifications in a purchasing project occurs regarding the mechanics or job specific ancillaries of the vehicle. The trend to under-spec often occurs regarding the ergonomics and crew comforts. Either of these extremes may be implemented with best intentions, but they can have disastrous results on the final effectiveness of the vehicle to carry out its intended purpose. Incorrect specifications on a new vehicle can also negatively affect the project timeline, unit cost, and running costs over a vehicle's life cycle. It is very important to be aware of this and give due consideration to creating appropriate specifications when deciding which vehicle characteristics will be essential and which will be desirable.

As stated, desirable characteristics are those which an organization would like to have on a new vehicle. However, the lack of them will not necessarily eliminate a vehicle from contention. Essential characteristics are exactly what they appear to be — they must be met by any vehicle supplier. They are those characteristics without which an offered vehicle will be rejected early during the purchasing process.

Your compliance matrix should be a comprehensive document. It is the first opportunity to exclude any manufacturers from contention — those whose vehicles may not even resemble what your company needs or intends to purchase. However, in my experience, the most important aspect of compiling a compliance matrix is to specify only what you actually require on your end product. Never over-spec or seek a higher specification for a vehicle than your fundamental requirements. Never specify technology that an RFI or reliable information has indicated is not readily available. This is most important with essential characteristics.

It may not be possible for any potential supplier to meet a compliance matrix requirement that is too technologically advanced and improperly researched, or worse, not researched. An item that is incorrectly specified as an essential requirement without proper preparation and research can jeopardize the entire project. In such circumstances, if no supplier can meet a stipulated essential characteristic — perhaps because the technology doesn't exist in your vehicle type — the result can be an unwieldy scramble to justify unrealistic compliance matrix specifications. You may also then have to change specifications downwards to keep a project alive.

This can cause further embarrassment or potentially kill the project if it appears that the required vehicle technology is not available, when in reality the organization stated a requirement beyond what was actually needed. The result is someone having to explain why an essential characteristic is no longer considered essential. Either way, the end result is negative for the project itself, and can reflect badly upon everyone involved and upon the professionalism of the organization.

**Never identify a vehicle characteristic that exceeds your fundamental requirements as being "essential." Use "desirable" characteristics to aim at higher specifications.**

If one manufacturer somehow manages to meet an essential requirement specified higher than what is actually needed, it may change the face of the entire purchasing process. Your organization may then be forced into buying a product that is lacking in a very significant number of areas, *solely* because there are no other providers able to meet a single requirement you designated as essential at the time you were seeking the vehicle. This may be the case even if you complete the total ratings for each contender, and other potential candidate vehicles are better overall.

Here is an extreme — and I hope, unlikely — example to highlight this point. Suppose four vehicles are competing for a contract. All vehicles are rated best to worst from 1 to 4. In order to give the manufacturers a particular goal in the project, you specify a power-to-weight ratio as essential, one that does not currently exist in this type of vehicle worldwide. Should the worst vehicle somehow manage to meet this single essential requirement, which the others do not, you may be forced into purchasing the worst overall vehicle.

This undesirable result occurs simply because the other three vehicles are eliminated by default for failing to meet an essential requirement. This problem may not be a binding one in the private sector, where the project evaluation can be overridden by a company owner or management board. However, in the public sector, purchasing a second-placed or third-placed vehicle, while failing to purchase the vehicle that meets an incorrectly specified essential characteristic — and to all appearances won the purchasing competition — can result in lengthy court battles and litigation.

On military purchases, an opinion is often held that more armor protection is always better. This view is, in fact, erroneous because the level of protection should be directly derived from the anticipated employment and role of the vehicle. We've previously seen this concern mentioned in the discussion of Stage 4, *Creating Your Upper Limits* — Serial No. 006. Increasing armor protection beyond what can be considered standard for a particular vehicle type can have a dramatic effect upon other vehicle characteristics. Again, these types of changes ultimately lead into R&D. The prototype definition and upper-limits definition for your purchasing project offer some constraint in this regard. However, they are not distributed to the competing vehicle suppliers or manufacturers and would not have legal standing if a dispute arose later.

**A project should never use the compliance matrix to try to give prospective suppliers a target to achieve. Any manufacturer worth their salt, if they exceed requirements in an important area, will let the project team know about it anyway.**

Compliance matrix specifications for your final vehicle will consist of stipulations such as "The vehicle must be wheeled and not use tracks." The matrix allows you to transfer to the potential suppliers some of the responsibility for the vehicle meeting your legal requirements. For example, in relation to braking systems, you may specify, "The vehicle must comply with European Union Council

Directive 71/320/EEC and all subsequent amendments." Or, in relation to compliance with an umbrella law, an essential characteristic might state, "The vehicle must comply with the Road Traffic Acts 1961 to 2011."

The compliance matrix will also specify how you confirm that a potential vehicle meets each requirement. A vehicle can meet a requirement through trials, which effectively means that your company will test the vehicle to ensure that it meets the required specification. In some cases, this testing may involve only a visual check — for example, is the steering wheel on the correct side? In other cases, the trial involves more extensive testing, such as the use of decellerometers to confirm braking force, or precisely engineered slopes to measure the ability to open heavy doors manually, etc.

Another option for ensuring that a vehicle meets a requirement is through Manufacturer Certification, which means manufacturers will be required to provide written certification that their vehicles meet the required specification. This approach obviously leaves manufacturers open to expensive litigation and reputational damage at a later date if the information they provide is incorrect. In my experience, this has never been an issue and is unlikely to be so if you are dealing with reputable manufacturers. The option of manufacturer's certification is usually used for very exacting scientific and legal standards where testing is expensive and time consuming.

*Figure 4-7 Sample Compliance Matrix*

| EQUIPMENT / VEHICLE PROJECT COMPLIANCE MATRIX | | | | | |
|---|---|---|---|---|---|
| **Serial** | **Description** | **Requirement** | **Requirement** Essential **(E)** Desirable **(D)** | **Qualification** Trials **(T)** Manufacturer Certification **(C)** | **Comment** |
| **1.0** | Speed | Vehicle must be capable of a sustained on-road speed of 110km/hr. | E | T | |
| **2.0** | Fuel | Equipment must be capable of operating for >12 hours at full capacity. (Or vehicle range > X Miles) | E | T | |
| **3.0** | Fording | Vehicle must be capable of a minimum unprepared fording depth of 0.5m | E | C | |
| **4.0** | Side Slope | Vehicle must be capable of operating on a minimum side slope of 20%. | E | C | |
| **5.0** | Temperature Range | Vehicle must be capable of operating in accordance with SAE Standard XXXX. | E | C | |
| **6.0** | Weight | Vehicle must have a GVM (Plated) no greater than 12000kg. | E | T | |
| **6.1** | | Vehicle must have a minimum payload capacity of > 1 ton. | E | C | |
| **7.0** | Cross Country | Vehicle to be fitted with tires suitable for both on-road and off-road use. | E | T | |
| **7.1** | | Vehicle must be capable of sustained operation off-road with no subsequently noticeable negative effect on brakes or steering. | E | T | |
| **7.2** | | Transfer case must facilitate constant 4x4 operation. | E | T | |
| **8.0** | Capacity | Vehicle must have the following minimum crew capacity: 1 x Drive | E | T | |

*continued*

| | | | | | |
|---|---|---|---|---|---|
| **8.1** | | Vehicle should have the following minimum crew capacity: 1 x Driver 2 x Passenger | D | T | |
| **9.0** | Engine | Vehicle must be fitted with a diesel engine generating a power:weight ratio of no less than 23:1. | E | T | |
| **9.1** | | Vehicle should be capable of using LPG. | D | T & C | |
| **9.2** | | If vehicle uses a direct drive forestry cutting disc, this must be capable of being manually disabled by a person outside the vehicle through the use of a readily accessible emergency button at the opposite side to the cutter. | E | T | |
| **9.3** | | Vehicle should be fitted with an engine block heater for starting in cold conditions. | D | T | |
| **10.0** | Emissions | Vehicle must meet EURO-III emissions standards. | E | C | |
| **10.1** | | Vehicle must meet EURO-IV emissions standards. | D | C | |
| **10.2** | | Vehicle exhaust efflux must be directed so as not to blow into the drivers cab. | E | T | |
| **11.0** | Idle | Vehicle must be capable of extended idling while operating a cutting disc for 10hrs in any 24hr period and 50hrs in any 5 day week. | E | T | |
| **12.0** | Steering | Vehicle steering must conform with Federal Motor Vehicle Safety Standard - FMVSS 203, FMVSS 204, FMVSS XXX, (Or 136, 70/311/EEC and XX/XXX/EEC as applicable) | E | C | |

*continued*

| EQUIPMENT / VEHICLE PROJECT COMPLIANCE MATRIX | | | | | |
|---|---|---|---|---|---|
| **Serial** | **Description** | **Requirement** | **Requirement** Essential **(E)** Desirable **(D)** | **Qualification** Trials **(T)** Manufacturer Certification **(C)** | **Comment** |
| **12.1** | | Steering box must be protected from damage due to external impact. | E | T | |
| **12.2** | | Vehicle will be fitted with effective power assisted steering. | E | T | |
| **12.3** | | Vehicle steering wheel will be located in the center of the cab. | E | T | |
| **13.0** | Brakes | Vehicle braking systems must conform with FMVSS 135, (71/320/EEC etc.) | E | C | |
| **13.1** | | Vehicle must be fitted with an anti-lock braking system. | E | T & C | |
| **13.2** | | ABS Sensors and cabling must be routed and protected so as to minimize likelihood of damage inside forests. | E | T | |
| **13.3** | | Braking system must be capable of holding vehicle on a 70% slope without pedal use by driver. | E | T | |
| **14.0** | Restraints | Vehicle must be fitted with restraints in accordance with FMVSS 202 (FMVSS XXX, 77/541/EEC etc.) | E | C | |
| **15.0** | Bodywork | Vehicle will be capable of undergoing an external cleaning process using high pressure water jets ( > 130Bar). | E | T | |
| **15.1** | | Vehicle will not have external recesses where water can collect and cause corrosion. | D | T | |
| **15.2** | | Interior floor and fittings must provide the maximum level of durability. | D | T | |
| **15.3** | | Exterior must be finished in the following color: XXXX | E | T | |

*continued*

| | | | | | |
|---|---|---|---|---|---|
| 15.4 | | Horizontal exterior surfaces which facilitate stepping must be coated in a non-slip finish. | D | T | |
| 16.0 | Maintenance Support | Manufacturer must provide a number (TBD) of spare parts catalogues available showing MTBF/MMBF, MTBCF, MTTR, LTTA. (COTS etc.) | E | T | |
| 16.1 | | Vehicle makes maximum use of non-unique or manufacturer licensed parts. | D | T | |
| 16.2 | | Spare Parts catalogue must list original equipment manufacturer (OEM) name and OEM part numbers. | E | T | |
| 16.3 | | An itemized price list must be available for vehicle spare parts supplied direct from the vehicle manufacturer for the current year - 2004. | D | T | |
| 16.4 | | A complete list of Special Tools and Test Equipment required to maintain the vehicle (If any) must be available. | D | T | |
| 16.5 | | Vehicle manuals must be in the English (Spanish, French etc.) language. | D | T | |
| 16.6 | | For no less than 20 years, manufacturer will provide maintenance support for the vehicles (Or equipment) on an 'as-requested' basis, using their dealer network in Alaska. In the event of parts' obsolescence, the vehicle manufacturer will recommend an alternative component. | D | C | |

*continued*

| EQUIPMENT / VEHICLE PROJECT COMPLIANCE MATRIX | | | | | |
|---|---|---|---|---|---|
| **Serial** | **Description** | **Requirement** | **Requirement** Essential **(E)** Desirable **(D)** | **Qualification** Trials **(T)** Manufacturer Certification **(C)** | **Comment** |
| **16.7** | | Vehicle should use an appropriate level of Mechanical Diagnostics. | D | T & C | |
| **16.8** | | Vehicle should be fitted with an easily removable belly cover to protect from damage due to impact with sawn logs and debris. | D | T | |
| **16.9** | | Vehicle should make maximum use of quick release couplings on hydraulic and fluid piping. | D | T | |
| **17.0** | Development History | Vehicle will have a proven in-service reliability record and not be a prototype (as defined by purchaser). | E | T | |
| **18.0** | Communications | Vehicle will be capable of providing a power supply for XXXX radio system and be fitted with a mounting in the drivers cab. | E | T | |
| **19.0** | Equipment | Vehicle must be fitted with an internally mounted hand-operated fire extinguisher which does not use HALON as the extinguishing agent. | E | T & C | |
| **19.1** | | Vehicle engine bay must be fitted with a Fire Detection and Suppression System (FDSS) which does not use HALON as the extinguishing agent. | E | T & C | |
| **19.2** | | Vehicle should be fitted with a self recovery winch certified in accordance with EN-14492/2010 etc. | D | T | |
| **19.3** | | Vehicle should be fitted with external white floodlights covering a 270 degree arc centred | E | T | |

*continued*

| | | | | | |
|---|---|---|---|---|---|
| 19.4 | | Vehicle will be fitted with a vehicle/operator monitoring system, with second-by-second recording of vehicle speed & rpm, and disc cutter rpm etc. | D | T | |
| 20.0 | Operational Functionality | Vehicle will be fitted with a towing hitch. | D | T | |
| 20.1 | | Vehicle will be fitted with a cold/warm air Air Conditioning system for the drivers cab. | D | T | |
| 20.2 | | The drivers cab will provide rollover protection in compliance with all state and federal law. | D | T | |
| 20.3 | | Vehicle cab design must minimize the danger from wooden projectiles due to disc cutter usage. | D | C | |
| 20.4 | | Vehicle should be fitted with a self-sealing fuel tank. | D | T & C | |
| 20.5 | | Vehicle must be fitted with a differential lock. | D | T | |
| 20.6 | | Vehicle should be fitted with an internal over-pressure and filtration system to provide clean air to the driver, for use in the vicinity of forest fires. | D | T & C | |
| 20.7 | | Vehicle should be fitted with side bars to allow external strap-on loads. | D | T | |
| 20.8 | | Vehicle cab must not be internally lockable. | D | T | |
| 20.9 | | All doors and windows must be fitted with waterproof seals. | D | T | |
| 20.a | | Vehicle should be fitted with run-flat tires. | D | T | |
| 20.b | | Cutting disc in stowage position must not protrude beyond the vehicle wing mirrors. | D | T | |
| 20.c | | Vehicle must be capable of operating with wheel chains. | D | T & C | |

*continued*

| EQUIPMENT / VEHICLE PROJECT COMPLIANCE MATRIX | | | | | |
|---|---|---|---|---|---|
| **Serial** | **Description** | **Requirement** | **Requirement** Essential **(E)** Desirable **(D)** | **Qualification** Trials **(T)** Manufacturer Certification **(C)** | **Comment** |
| **20.d** | | Vehicle must comply with the following U.S. State and Federal (Or European as applicable) Regulations: 49 CFR Part 571 etc. | E | C | |
| **20.e** | | Vehicle must comply with the following SAE Standards: XXXX | E | C | |
| **21.0** | Ergonomics | Vehicle will facilitate safe operation by persons within the 5th to 95th percentile. | D | T | |
| **21.1** | | Vehicle must be fitted with a driver seat capable of swivel, and also adjustment in a minimum of 2 axis. | D | T | |
| **22.0** | H & S | Vehicle will comply with the Safety, Health and Welfare At Work Act 2005, the Safety Health and Welfare At Work Regulations 2007 (Or US/EU Federal/State as applicable) | E | C | |
| **22.1** | | At least one (1) exit point will be capable of being opened when the vehicle is resting on its roof or side. | E | T & C | |
| **22.2** | | Ride quality will ensure that personnel in the vehicle do not suffer undue motion sickness. | D | T | |
| **22.3** | | Radio will have an external antennae mounting point as specified by the manufacturer of the following aerial: XXXX | E | C | |
| **22.4** | | Vehicle must be free from asbestos. | E | C | |

*continued*

| | | | | | |
|---|---|---|---|---|---|
| 23.0 | Protection | Vehicle cab should be fitted with an internal liner to minimise likelihood of penetration and driver injury from cutter debris. | E | T & C | |
| 24.0 | Recovery | Vehicle should be fitted with a trickle charge socket and transformer, with 110V AC input, and a 15m charge cable. | E | T | |
| 24.1 | | Vehicle will be capable of carrying – shovel, 5ft 2 person crescent ground felling saw, 1 person crescent ground bucking saw, raker gauges, crow-bar. | D | T | |
| 24.2 | | Vehicle should be fitted with 5m jump cables. | D | T | |
| 24.3 | | Vehicle should be fitted with a breakdown kit including 2 sets of roadside hazard lights. | D | T | |
| 24.4 | | Vehicle must be fitted with an externally mounted spare wheel. | D | T | |
| 24.5 | | Vehicle transmission must facilitate towing. | D | T | |
| 25.0 | Electrical | All vehicle Electrical components and connectors situated below – fording depth + 50% – must be IP57 standard. | D | C | |
| 25.1 | | External mirrors must be fitted with electrical de-icing. | D | T | |
| 25.2 | | Driver warning and gauge lights must be capable of being dimmed. | D | T | |
| 25.3 | | Internal white lights must be capable of being dimmed by the driver. | D | T | |
| 25.4 | | Vehicle must be fitted with flashing amber hazard roof lights capable of being turned on/off by a switch inside the cab. | D | T | |

*continued*

| EQUIPMENT / VEHICLE PROJECT COMPLIANCE MATRIX | | | | | |
|---|---|---|---|---|---|
| **Serial** | **Description** | **Requirement** | **Requirement** Essential **(E)** Desirable **(D)** | **Qualification** Trials **(T)** Manufacturer Certification **(C)** | **Comment** |
| **25.5** | | Vehicle must be fitted with non-flashing lights along the cutter arm. | D | C | |
| **25.6** | | Vehicle/Equipment should have a switch on/off fast-idle capability for battery charging. | D | T | |
| **25.7** | | Final electrical power budget must allow for up to 25% future growth. | D | T | |
| **25.8** | | All vehicle diagnostic plug connections must be easily accessible on a single mount. | D | T | |
| **25.9** | | All dashboard warnings must be amber or red. | E | T | |
| **25.a** | | Vehicle must be fitted with variable speed wipers front and rear, and a windscreen washer system. | E | T | |
| **25.b** | | Minor electrical faults must not disable the vehicle. | E | T | |
| **25.c** | | Vehicle batteries will have sufficient ventilation to prevent over-heating. | E | T | |
| **25.d** | | Vehicle batteries should be sealed and maintenance free. | E | T | |
| **25.e** | | Vehicle batteries must be in a location that is readily accessible. | D | T | |
| **25.f** | | Vehicle must be fitted with pop-type fuses. | D | T | |

*continued*

| | | | | | |
|---|---|---|---|---|---|
| 25.g | | Vehicle must be fitted with the following gauges/lights:<br>▪ Speedometer<br>▪ Tachometer<br>▪ Output Drive Tachometer<br>▪ Odometer<br>▪ Fuel gauge<br>▪ Battery condition indicator<br>▪ Engine coolant temperature gauge<br>▪ Low coolant level warning light<br>▪ Low engine oil pressure warning light<br>▪ Low battery voltage warning light<br>▪ Etc. | D | T | |

Vehicles will often have been tested independently during the original design process in order for them to be legally marketable by the manufacturer. Examples include braking standards and steering standards.

In some cases, both trials and manufacturer certification will be required to confirm that a vehicle meets the specified requirement. An example of this is Serial 19.0, in Figure 4-7, where a visual check during trials will confirm there is an extinguisher present and the manufacturer's certification will confirm the nature of the extinguishing agent. In many cases, the vehicle manufacturer will be waiting on the OEM of various components to provide the required certifications. Therefore, when originally going to market to seek vehicles, patience must be exercised and sufficient time allowed for the vehicle manufacturer to review your issued compliance matrix and gather the required certifications prior to proceeding with the purchasing project.

Figure 4-7 shows a sample compliance matrix. This matrix is by no means exhaustive —the areas for compliance are in many ways specific to each individual project. This matrix, however, is largely project neutral; it contains specifications that can be used as a foundation on many projects.

A full compliance matrix for a new vehicle purchase program may often be much longer than the sample in Figure 4-7. It can run tens of pages or more, depending upon how exact you — the purchaser — wish to be in your vehicle specification. The compliance matrix must be as comprehensive as possible, because it should also be used to create the practical check sheets for eliminating vehicles prior to and during the trials process.

When specifications call for legally binding test standards, the formal certification of the manufacturer or an independent company is required. If time, budget, and facilities allow, it is desirable to test these standards independently during trials, but often not required unless questions arise from the trials regarding the functionality of specific vehicle systems.

Sometimes significant questions can arise from the trials process surrounding particular aspects of a vehicle. In these cases, it can be agreed with a manufacturer that should their vehicle be selected for purchase, required certification will be provided by an independent third party. Dependent upon vehicle type and the requirements of your individual organization, various items can be added to or removed from the example compliance matrix in Figure 4-7.

## Stage 7: Initial Review of Potential Vehicles

The previous stages focused primarily upon the preparatory work required for a fleet purchasing project. — Creating an RFI allows your organization to be certain that the type of technology you intend to purchase is actually available in the market place. Defining your prototype definition and upper limits gives your project team clear boundaries for the project. Defining your compliance matrix lets your organization set out clearly the specifications and technical requirements for your potential new vehicles.

At this point you will have sought submissions from potential suppliers and received back documentation and technical details of the vehicles they can offer. Alternatively, your project team will have sought out such information and compiled a list of potential vehicles for further investigation and evaluation.

Stage 7 in the evaluation of supplier submissions generally consists of a paper-based review and initial elimination of candidates. This can be a valuable and cost-effective means to reduce the number of potential vehicles prior to the more expensive and time consuming vehicle trials. Ideally you do not want to be conducting practical trials and evaluations on more than three or four vehicles because it will take too long and cost too much. This is especially the case on projects with a larger financial value.

The fundamental purpose of this stage is twofold. First, it can weed out any completely unsuitable supplier submissions from the competition, as determined by the overall requirements. Second, it can also reduce, to a manageable quantity, the number of vehicles which proceed to the next stage of evaluation.

The first step in the initial evaluation process is to decide upon the maximum number of vehicles that can be selected to proceed to the trials stage of the purchasing process. This decision is largely dependent upon your company's capacity to conduct trials, the timeline available for the completion of those trials, and the resources available for practical trials. For example, if you have sufficient personnel and resources to test only two vehicles at the same time, and you have only a small window of a few days or weeks for trials, then running the vehicle trials process for more than two vehicles will not achieve accurate or detailed evaluation results. If the number of vehicles going forward to your practical trials is being limited prior to the initial review of submissions from suppliers, this should be noted to the manufacturers, and a marking scheme should be established for elimination. The marking scheme does not have to be communicated to the suppliers. However, depending upon your country's laws, it may have to be available in response to later Freedom of Information requests.

The actual decision regarding which vehicle submissions are appropriate to continue to trials is based upon a cross checking of the manufacturers' submissions against the compliance matrix. Vehicles can easily be eliminated based upon failure to meet any of the essential requirements.

There is a possibility that all vehicles fail to meet at least one essential requirement. In this case, there may not be a vehicle available in the marketplace to fulfil your specifications. If this failure is due to having contradictory essential requirements in your compliance matrix, or a requirement set higher than what is available, then it is an error in the matrix. Otherwise, you may simply be unable to purchase the required vehicle at this time — the project may need to be stalled or halted until the market matures. However, this problem should have been established previously with the use of an RFI.

At this stage in the purchasing process, you may also wish to receive oral presentations from each manufacturer regarding their available vehicle and technology. The manufacturer's willingness to make these presentations may depend upon the scale of your potential fleet purchase. At this point, potential vehicles will be dismissed or accepted for further evaluation, according to their ability to meet the essential requirements in the compliance matrix. However, if this step still results in too many vehicles remaining, you can further reduce the list of vehicles going forward to practical trials through an assessment of how highly they are rated against desirable characteristics in the compliance matrix.

At this juncture, presentations from the manufacturers to your project team allow an opportunity to ask questions, receive clarifications, and gain an impression of the nature of the companies. If vehicles are eliminated solely using essential requirements, then no marking scheme is required yet. This is because vehicles are simply deemed to be in accordance with the tender requirements, based upon the compliance matrix or they are noncompliant and eliminated. If, however, too many vehicles are compliant and can potentially proceed to the next stage — which is vehicle trials — then marking or scoring desirable characteristics can be used to eliminate some more contenders.

For any purchase of unique equipment, there is a significant risk from the purchaser's perspective. This risk is that a manufacturer may use your project to effectively try to field test or prove one or more new vehicle concepts. This attempt by manufacturers can present an acute risk when the purchasing project involves a small number of vehicles or the quantity of vehicles for modification is smaller. The risk is more often experienced in military or specialized environments, when equipment is less Commercial Off The Shelf (COTS). We must constantly be aware of this risk from the outset of any project and factor it into our decision-making process regarding any manufacturer proposals for us to purchase new technology or systems.

The benefit-cost of using your project to test something new may swing in your — the purchaser's — favor to negate this risk. This depends upon the potential profits to be made by the vehicle supplier or manufacturer from your business with them versus the potential profits to be made from other sales of the new equipment or concept, if it is successful. If you are a large or long-time customer of the manufacturer, it is not in their interest to risk alienating your business to prove a new concept. This fact alone can give you more confidence when considering such options. It is often better for a manufacturer to risk field testing new technology on vehicles being produced for a smaller one-time customer, where there is less present and future business to lose.

Once this initial review has been completed, and the number of vehicles to proceed to practical trials and evaluation has been established, the manufacturers or suppliers can be notified of your decision, and we can proceed to practical vehicle testing.

### Stage 8: Trials of Potential Vehicles

Trials make up the largest and most important aspect of the entire vehicle procurement process. Dependant upon the size and financial value of an overall purchase, the scale of your trials can vary greatly. Your trials may involve simply testing vehicles at a location specified by each manufacturer over a number of days; on the opposite end of the scale, your trials may involve the outright purchase of vehicles and detailed evaluation over a number of years at various locations. This testing can take place in various climates worldwide and ultimately culminate in testing to destruction.

The more time you make available for evaluating potential vehicles, the more successful the trials process will be. More time for testing allows more detailed and in-depth evaluations to be completed, whereas time pressure during testing can limit the quality and scale of the assessments.

It is desirable to test all vehicles in the same locations, as specified by you the purchaser. Although this may result in your organization incurring costs associated with transporting each manufacturer's vehicles from factories to the required locations, it allows a more balanced evaluation of the various vehicles. If your project value and potential subsequent fleet purchase value is high enough, however, manufacturers will often provide vehicles to your location at their own expense.

Testing at the manufacturer's individual locations could leave you unable to test the details of all contenders to the same extent — for example, there may be differences in what testing equipment is available at each manufacturers facilities. Differences in climate and environmental conditions at different locations can also contribute to a non-standardized testing environment, which is not desirable for our trials. As purchasers we prefer to have the manufacturer's vehicles come to us for as much of the practical testing as our own facilities and resources will allow.

A separate factor is that our facilities and resources may not be sufficient for the testing we would ideally like to undertake. We may, therefore, have to receive competing vehicles at our own location, and then transport them to other test sites during the course of the trials. These other test sites can vary in type and size depending upon the scale of our trials and the nature of the individual tests being carried out. For example, the vehicles may be driven a few miles/km to a local weighing area that we have paid to use for a few hours in order to confirm the net weights of the vehicles.

Or, suppose your company needs a vehicle to operate successfully in both arctic and desert conditions. There is no substitute for actually testing them on site in such areas. Therefore, we may be willing to incur significant costs to transport vehicles to different potential operating environments in order to undertake comprehensive operational evaluations. Unfortunately this option is not always available due to cost or simple practicalities.

However, in the civilian logistics field, such testing on potential new vehicles is almost never undertaken — it can be reliably avoided using the option in our compliance matrix for manufacturer certification. In the specialist or military fields, such testing can also be reliably avoided — in most instances — by requiring manufacturers to certify their vehicle in accordance with the relevant standards. The extreme limits of the requirements envelope are thus facilitated through certification by the manufacturer in accordance with internationally recognized standards; we avoid the time, expense, and complexity of testing in various operating environments.

That said, there is no substitute for the physical testing of new vehicles functionality and performance in different operating environments. It allows us to reliably determine the positive and negative aspects of our potential new vehicles in the actual locations they will operate. We can then incorporate this information into both our life cycle cost projections and crucially — our evaluations of future reliability and maintenance implications. As you can imagine, an articulated tractor unit operating daily in an arctic environment will have very different and unique operating challenges to one that is operating in a desert environment on the equator.

The vehicle trials should encompass a number of different evaluation areas, as described below:

### Mechanical Evaluation

**Mechanical Examination and Tests of the mechanical properties of the potential vehicles, their likely ongoing maintenance requirements, and any associated technical considerations.**

The mechanical evaluation you need to carry out physically on each vehicle at this stage of the process allocates points based upon the evaluation of two areas: Technical Merit and Technical Performance is the first; Ease of Maintenance is the second. Preferably you will use your own in-house maintenance staff for this process because they will ultimately be the persons maintaining the new vehicles. If necessary, you should seek assistance from your maintenance contractors who will be involved in the ongoing maintenance of the fleet. Obviously there are costs involved in this step. In fact, your financial constraints will have played a significant role at the outset of the purchasing process in determining the scale of these trials. The points are allocated as per the table in Figure 4-8.

Suppose the total points to be allocated for this project are 3000, as previously discussed in the section *Forming A Project Team and Points Allocation.* The mechanical trials will perhaps be allocated 33% or 1000 points from this total. These 1000 points should be split between the two mechanical evaluation areas — [Technical Merit and Technical Performance] [Ease of Maintenance] — for the practical trials, usually in a 50/50 split.

| Points | MECHANICAL EVALUATION POINTS ALLOCATION | | |
|---|---|---|---|
| 5 | V-EASY TASK | OR | ALWAYS EXCEEDS REQUIREMENTS |
| 4 | EASY TASK | OR | USUALLY EXCEEDS REQUIREMENTS |
| 3 | SATISFACTORY TASK | OR | MEETS REQUIREMENTS |
| 2 | AWKWARD TASK | OR | SOMETIMES FAILS TO MEET REQUIREMENTS |
| 1 | VERY AWKWARD | OR | OFTEN FAILS TO MEET REQUIREMENTS |
| 0 | UNWORKABLE | OR | ALWAYS FAILS TO MEET REQUIREMENTS |

***Figure 4-8, Mechanical Evaluation Points Allocation,***

The assessment of Technical Merit and Technical Performance for each potential vehicle involves a physical examination of the vehicle, without the manufacturers present. Usually your mechanical personnel will be split into teams of two in order to carry out this evaluation. Each team will be allocated a number of items to evaluate. To verify conformity with the technical details contained in the compliance matrix, each team will be given a selection of the overall technical requirements, which they have to evaluate on every vehicle. To do this, each team will use a standardized evaluation matrix, an example of which is shown in Figure 4-9. They will allocate points to their particular areas using the point allocations in Figure 4-8. While Figure 4-9 only lists 8 items for assessment it is simply an example. Your technical merit and performance evaluation sheets will often extend to hundreds or thousands of individual items.

The reason that each team is not allocated a specific vehicle to assess all areas is that the marks for each specific item must be assessed in comparison to the other competing vehicles. By giving each team a specific list of items which they will assess across all vehicles, we negate the possibility of differences in the personal perceptions of team members affecting the final comparison of the vehicles. For example, if one team is marking more generously than another team, it will not matter because that team will mark all vehicles more generously; the distribution of marks to each vehicle will still rate them accurately against each other.

When your teams are approaching this part of

the evaluation, the standard mark allocated to an individual item for assessment does not begin at 0 or 5, and then gain or lose marks from this level during the evaluation. Instead, the mark at the outset of the evaluation for every individual item, such as “All doors will be fitted with locking devices,” should be 3 points to represent Meets Requirements. Any variation to this item that is noted during the evaluation will thus be deemed better or worse than requirements. The item will then be allocated a higher or lower score based upon the assessment of the individual team responsible for that item.

The manufacturer should not be present for this evaluation so that all vehicles can be examined impartially without interference. However, they should be easily contactable for any queries or clarifications, as required. Any technical clarifications received verbally from a manufacturer should be followed up in writing as soon as possible, preferably on the same day — email provides a convenient and quick method.

Figure 4-9 shows an example of how a Technical Merit and Technical Performance evaluation sheet is compiled and used. A full version will cover many more of the compliance matrix topics, in addition to any other topics which you wish to evaluate. These evaluation sheets are NOT limited to the items contained in the compliance matrix. You may have become aware of additional positive or negative technical features during the initial vehicle evaluations — areas for which you now want to award or deduct points. Be sure that all competing vehicles are assessed for each individual item so

| Technical Merit and Performance Evaluation | | | |
|---|---|---|---|
| Serial | Aspect | Marks | Remarks |
| 1. | Vehicle will be fueled and weighed. | Not Marked | |
| 2. | Vehicle will complete national DOE/NCT test. | Not Marked | |
| Engine/Powerpack | | | |
| 3. | Have a diesel engine generating a minimum power:weight ratio of 23:1 hp/t (DIN). | 2 | Power to Weight ratio 21:1 |
| 4. | Exhaust Gases should be directed so as not to raise dust. | 3 | Meets Requirements |
| 5. | Escaping smoke should be Low/Gray by visual assessment. | 3 | Meets Requirements |
| 6. | Vehicle is fitted with a high-idle facility. | 0 | Not fitted |
| 7. | Vehicle engine safe mode can be disabled manually. | 0 | Cannot be overridden |
| 8. | Alternator is located for sufficient cooling | 4 | Alternator is liquid cooled |

***Figure 4-9, Technical Merit and Performance Evaluation Table***

that the evaluation is fair and accurate.

The columns for Marks and Remarks are completed during the evaluation of the vehicles. They have been filled in here for example only.

To summarize: in this example we have allocated 3000 points in total to the project. Of this, 1000 points have been allocated to the Mechanical Trials. In turn, of this 1000 points, 50% (or 500 of the original 3000 points) goes towards Technical Merit and Performance.

Within the context of Figure 4-9, the most marks a vehicle can achieve for technical merit and performance — the maximum marks attainable — is 30, which would be achieved if every item was awarded 5 marks. A score of 30 marks represents 100% of the project points in this area, or 500 points. However, the vehicle evaluated in Figure 4-9 was awarded only 12 marks, or 40% of the 30 marks available. Therefore, this vehicle would be awarded 40% of the available 500 project points, or 200 points. (In short 12 of 30 evaluation marks is equivalent to 200 of 500 project points.)

Remember that this is a working sheet. Both the marks and the remarks to justify them can be entered as the evaluation proceeds. Then, if necessary, they can be reviewed and amended upwards or downwards during later group discussions. Most of the items in this evaluation will be assessed visually or by checking specifications for the vehicle. Some items may require more comprehensive detailed testing.

The second area of evaluation, Ease of Maintenance, involves physically carrying out maintenance tasks on each vehicle and marking the results. This evaluation is carried out with the manufacturer's representative present because your personnel are unlikely to be familiar with the process for carrying out every maintenance task on every unfamiliar vehicle in the most efficient way possible.

Each maintenance task should be recorded for how easy it is to complete, using the 0 – 5 point marking scheme from Figure 4-8. Each task should also be recorded for time and estimated time taken. The time is the actual time it took to complete the individual task during your evaluation. The estimated time is the approximate time the task is deemed likely to take once maintenance personnel are fully familiar with it. As many and varied tasks should be carried out as possible within the constraints of the overall time available for the Ease of Maintenance trials.

Tasks such as removal and re-fit of engine, transmission, starter motor, alternator, steering box, wheel hub, primary ECU, and drive shaft are all important and should be physically completed during the trials. A major service should also be completed to allow evaluation of the time and costs required. As before, individual team members should be allotted specific tasks for them to carry out on all of the trial vehicles, thus allowing the estimated time and comparisons to be more accurate. For example, if there are three vehicles for evaluation, the same mechanic should remove and refit the starter motor on all three, allowing an accurate comparison of the task across vehicles. Another mechanic can remove and refit the alternator on all vehicles, and so on.

| Ease of Maintenance and Technical Support | | | |
|---|---|---|---|
| Serial | Aspect | Marks | Remarks |
| Removal and Replacement of Components | | | |
| 1. | Conduct a vehicle service which involves an oil change. | 4 | Very easy access to all consumable parts. |
| 2. | Remove and Re-Fit/Replace Engine | 3 | Meets Requirements |
| 3. | Remove and Re-Fit/Replace Alternator | 3 | Meets Requirements |
| 4. | Remove and Re-Fit/Replace Starter Motor | 3 | Meets Requirements |
| | Etc… | | |
| Technical Support | | | |
| 5. | Manuals are legible in the English Language | 3 | Meets Requirements |

***Figure 4-10, Ease of Maintenance and Technical Support Table***

**Although these tasks may seem routine and unnecessary to test, there are often a huge variety of differences in the execution of even the most seemingly routine tasks across vehicles from different manufacturers. On one vehicle I experienced very early in my career, the replacement of a starter motor required the removal of the front grill and radiator in order to gain access, thus significantly extending the time taken for the task in comparison to other similar vehicles. Examples such as this are commonplace.**

Figure 4-10 shows an example of a completed Ease of Maintenance evaluation sheet.

**Different people evaluating the same task on different vehicles will not give an accurate comparison of the vehicles.**

Figure 4-10 provides a short example. The list of items to be included in your full Ease of Maintenance and Technical Support evaluation, depends entirely upon your time constraints and capacity during the trials to physically carry out as many maintenance tasks as possible. The items evaluated under the Technical Support heading will reflect your specific maintenance support requirements. The overall technical support evaluation will include an evaluation of COTS parts usage, lead times and costs, as explained later in this chapter.

## Mobility Examination

**Mobility — Examinations and tests of the driving and mobility characteristics of potential vehicles.**

Mobility trials evaluate the capabilities of each vehicle to carry out the functional and operational tasks required after purchase to the highest possible standards. Depending upon the nature of the vehicles or equipment you are potentially purchasing, and also upon the previous experience of your personnel with such vehicles, your staff may require training prior to the beginning of the mobility trials. In most countries, your personnel cannot take a trial vehicle out onto a public road if they do not have a driving license; similarly, they may be also unable to accurately operate and evaluate the trial vehicles or equipment without first becoming familiar with their unique features. As needed, you may have to make arrangements with the competing manufacturers to undergo this training a few days or weeks prior to the mobility trials. The exact nature and extent of such training should be determined by the manufacturers — they are the ones most familiar with the operation of their vehicles.

Mobility testing should begin by evaluating the basic driving characteristics of each vehicle, such as vehicle handling over different types of terrain, fuel usage and range over different types of terrain, and maneuverability. Mobility trials can also incorporate a general series of tests such as slalom runs, high speed driving, off-road driving, and other exercises appropriate to the desired end use of the vehicles.

For example, if you intend to purchase a logging vehicle that must operate reliably in swamps and over soft ground, there is little benefit in testing it only on hard-surfaced roads. The vehicles must be tested in the types of environment where they will ultimately operate after being purchased, or as close to this environment as realistically possible within your budget and capabilities. If it is not possible to simulate or reliably test the vehicles in their final operating environment, you must seek details from the manufacturer about other customers who operate the vehicle in the relevant environment. In addition, details of the type of testing the manufacturers have completed during the vehicle development can provide further peace of mind that if you choose to purchase a particular vehicle, it will prove effective in its future role.

The mobility evaluation involves driving and using the vehicles during the trials over the available time period, both on and off road. This usage of the vehicles and the associated assessment can take place either prior to or after the mechanical trials. However, it is not viable for both trials to take place at the same time because the vehicles will be required full-time for the mechanical assessments. In the case of a military vehicle, which throughout its lifetime will be required to operate overseas in various climates and both rural or urban environments, there is no substitute for carrying out mobility evaluations in-situ at a variety of locations. As always, this testing is dependent upon both financing and the total time available for trials. However, it has clear long-term financial benefits, significantly reducing the possibility of unforeseen operational issues after purchase.

It is important to ensure that your mobility trials are tightly defined at the outset — both their objectives and the total time permitted for them to reach their conclusion. Such planning can prevent them from extending over an inordinate period of time. There is a general difference in the purpose of the mechanical trials vs. the mobility trials. The mechanical trials assess in detail the individual technical features of each vehicle and evaluate individual tasks. The mobility tests during the mobility trials are less to evaluate the specific test, and more to facilitate as much varied usage of the vehicles as possible, enabling an overall understanding of the strengths, weaknesses, and functionality of each competing vehicle. Your personnel who are evaluating the vehicles can give an informed comparison of the trial vehicles. The trials should include high and low speed driving throughout the gear range and the full extent of each vehicle's capabilities.

Evaluation should also be carried out on the practicalities of using the vehicles in their intended final role. In the case of a military vehicle, you will want to evaluate such matters as whether or not the vehicles have sufficient physical space to carry all required mission specific equipment. In the case of civilian vehicles used for long distance haulage, evaluate whether they have simple necessities such as sleeper cabs, if required. The list of areas for evaluation can be as long or as short as you wish; it is limited only by the nature of your particular vehicle requirements.

## Support Examination

**Support — Examination of the spare parts, technical support, and after sales support available for potential vehicles.**

The support aspect of the vehicle trials ensures that any vehicle purchased fits in with your envisioned support requirements and capabilities.

Purchasing a vehicle that makes maximum use of COTS — Commercial Off The Shelf — parts can provide significant advantages in relation to reducing long-term costs and minimizing the spare parts LTTA — Lead Time To Availability. Military readers should note that COTS parts can give significant advantages in vehicle operations, but MOTS — Military Off The Shelf — parts do not give comparable benefits — the costs for military parts will always remain higher. However, MOTS parts are preferable to unique vehicle-specific or manufacturer-specific parts.

| Manufacturer COTS Spares Availability | |
|---|---|
| Number of Non-COTS or Manufacturer Licensed Vehicle Parts (NC) | 142 |
| Number of COTS Vehicle Parts (C) | 486 |
| Total Number of Parts Listed (T) | 628 |
| % of Total Parts Potentially Available COTS *(100/T)*C* | 77% |

***Figure 4-11, Evaluation Table for Manufacturer COTS Availability***

To evaluate spare parts, there are three key areas to assess during the purchasing trials: COTS usage, LTTA, and Cost.

Start by requiring each manufacturer to submit a comprehensive list of vehicle spare parts. This list should state the following for each individual spare part:

1. **Part Name**
2. **Part Number**
3. **OEM — Original Equipment Manufacturer — Company Name**
4. **OEM Part Number**
5. **Current Recommended Retail Price (For COTS parts)**
6. **COTS, MOTS, or Vehicle Manufacturer Specific component**
7. **LTTA**

There is often quite a bit of work involved for potential vehicle suppliers in compiling this list. As such, they need to be informed of this requirement as early as possible, preferably prior to the initial vehicle evaluation, so they cannot say there was insufficient time to prepare the data. Realize that unless the potential sales from your upcoming vehicle purchase are going to be significant, some manufacturers or suppliers may be unwilling to commit the necessary time and resources to the creation and release of this data. In the vast majority of cases, you can at least expect this list to be released as a confidential document.

Manufacturer specific components can be those created in-house by the vehicle manufacturer; they may also be those parts which the OEM has contractually licensed only that vehicle manufacturer to sell. In either case, the vehicle manufacturer is the only supplier. You will be tied into their LTTA and parts price for potentially the lifetime of the vehicle. This is not necessarily a deal breaker, but is a consideration to be aware of when choosing which vehicle to purchase.

The evaluation of COTS parts consists of determining the quantity of vehicle parts which are COTS versus the quantity which are specific to the vehicle manufacturer. The list in Figure 4-11 is used to establish this percentage. Once this list is completed for every vehicle remaining in your trials, the COTS percentage can be compared across all competing vehicle suppliers/manufacturers, and points awarded accordingly. The numeric values in the right-hand column are entered during the trials evaluation, using the spare parts list submitted by the manufacturers.

Now that we have evaluated COTS availability, we can move on to evaluating the average LTTA of the spare parts for each vehicle.

Each manufacturer should submit their list of spare parts with the estimated LTTA for each component. As with any aspect of the trials, if something is not present or a manufacturer fails to submit required information, they will normally receive a score of 0% for that area. It is not possible to award points for an area which you cannot evaluate!

These spare parts lists can be reviewed to calculate an average overall LTTA for each manufacturer's vehicle. The percentage awarded is based upon the maximum number of weeks which you are willing to accept as a lead time for any component.

Suppose you determine that an LTTA of over 26 weeks (6 months) is unacceptable for any component. Then an average LTTA of 26 weeks is equal to 0% score in your marking scheme. An LTTA of 0 weeks is always equal to 100%. In the following example, our average LTTA is calculated as 4.74 weeks. Using our maximum acceptable value of 26 weeks, awards a score of 82% to this vehicle:

$$[(26 \text{ weeks} - 4.74 \text{ weeks})(100/26 \text{ weeks})] = 81.77\%$$

Although this is a very simple method of comparing manufacturer spare parts' LTTA, it gives a valid comparative result. To see how the figures are established, see Figure 4-12.

The LTTA evaluation can be expanded significantly and made much more useful and detailed, by sub-dividing like-for-like components into groupings or sub-assemblies and evaluating these individually. For example, we would expect a much shorter LTTA on routine service components than on major assemblies. As such, we can mark different groupings of components using a different LTTA value for 0% as we wish. It is important to ensure that the same components are being marked on all vehicles in the same way.

| Manufacturer LTTA | |
|---|---|
| Number of Spare Parts Submitted with LTTA (N) | 628 |
| Total LTTA (Weeks) for All Parts (TL) | 2978 |
| Average LTTA (Weeks) (Avg) (TL/N) | 4.74 Weeks |
| Awarded % | 81.77% |

***Figure 4-12, Manufacturer LTTA***

Alternatively, establish your own secondary spare parts list for each manufacturer, based upon their original lists. You can then use these secondary lists to evaluate only those components that are common to all vehicles and eliminate others that are unique to individual vehicles. If you so wish, you can also separate the small number of components with very large LTTA or very small LTTA, which can skew the average. If taking this approach, however, it is very important to make sure that your results are not becoming inaccurate or not a true reflection of the LTTA situation. By eliminating unique components from your evaluation that may have a major effect upon later maintenance times, you can create a false result that will be detrimental to the project and the company in the long term.

The average cost of parts for each manufacturer also requires evaluation. This evaluation can be carried out in a general way by selecting all common components across all manufacturers. Such components would include those present on all vehicles, such as service brake pads, transmission, alternator, and starter motor, for which you have received prices from each contender. The manufacturer with the least expensive average component price is allocated the highest marks because it will cost less to purchase these parts.

You can also conduct this evaluation in a much more detailed manner by grouping similar components across manufacturers (e.g., Driveline, Bodywork, Internal Trim) and marking each group's average price individually as described above. This approach will give a more thorough analysis of Cost. Note that Cost of Spare Parts should not be given too much weight in your analysis. Prices can fluctuate over time; if the chosen vehicle has high COTS parts usage, your organization will likely buy the majority of spares from suppliers other than the vehicle manufacturer once the vehicle warranty expires.

When scoring vehicles and allotting points based upon the average price of their spare parts, another important consideration is that the MTTF (Mean Time To Failure) and MTBF (Mean Time Between Failure) of these parts can have a dramatic effect on their value for money. The cheapest spares do not necessarily represent the best value for money. This point is discussed in the second half of this book. For now, be aware that you should also consider failure rates in your spare parts cost analysis. This analysis may not be possible; however, if you do not have sufficient recorded analytical data to do so.

Another consideration is the Technical Support evaluation for each competing vehicle. This evaluation is not scored in the traditional sense. It essentially involves questioning each manufacturer regarding their current customer support structures worldwide. Most manufacturers will offer a variety of service and maintenance contracts to potential customers; the benefit-cost of these requires careful evaluation. For the purpose of your vehicle selection — unless such a maintenance contract is part of your tender or purchasing requirements — the primary concern is simply whether or not these facilities are available and how extensive is the manufacturer's after-sales support structure. For example, do they provide direct overseas service support to your vehicles? It may be possible to ask other fleet operators using each competing vehicle type about their experiences with the vehicle suppliers or manufacturers.

You should also evaluate the quality of the manufacturer's technical documentation if you intend to maintain your own vehicles in-house. Any minor customer-specific requirements you have, such as all manuals being printed on waterproof and tear proof paper to ensure longevity when in use, should also be communicated to each manufacturer at this time in order to ascertain their willingness and capacity to fulfill the requirement.

## Stage 9: Vehicle Selection

If you have completed the stages up to this point in a comprehensive and detailed manner, then actual vehicle selection should be the most straightforward aspect of the process. Consequently, this is perhaps a surprisingly short section of this book.

Prospective purchases should have been whittled down since the Initial Review of Potential Vehicles. By now, a final successful candidate should be clearly established from the vehicle trials and scoring matrices. Should the trials have resulted in more than one viable candidate vehicle, cost can be used to determine the final successful choice. However, the danger in vehicle selection arises when outside influences override the results of the vehicle trials. The most likely overriding factor is cost. The dangers to a project of cost being given an overarching influence have already been discussed, perhaps ad nauseum! This emphasis is only because the point is so important and worthy of reiterating.

However, other factors that can affect the final decision are alternative strengths and weaknesses of the competing vehicles in mission or role-specific areas. The final purchase decision should always be fundamentally based upon the vehicle that is most mechanically sound. All other issues should be reviewed secondary to that.

When the final decision is being made regarding which vehicle to purchase, I would go so far as to consider the allocation of a veto to the project team representative responsible for the Mechanical Performance evaluation. If a vehicle that you ultimately purchase had more issues mechanically during the selection process than other vehicles, your organization may be looking at addressing and attempting to rectify technical problems post delivery over a number of months or years. You are potentially looking at significant future life-cycle costs and failure-related introductory expenses. The promises of a manufacturer to resolve technical issues discovered during your trials must be carefully considered with respect to the timeline to reach the required resolutions. Although you should not fear purchasing a vehicle that requires some modifications, it is an expensive lesson for you to learn if you do not fully consider all implications.

Having selected for purchase the vehicle that scores best overall from all the evaluations and that is suitably mechanically sound, any other considerations can be prioritized according to the role profile envisioned for the vehicle and the organization's specific priorities for this vehicle after purchase. If the mechanical soundness of the design is given priority in the final vehicle selection, it gives a solid fundamental base from which to address other considerations. Any other issues can now be reviewed and addressed as the project progresses into the Design Phase that follows the vehicle selection.

Depending upon the uniqueness or specialist nature of your vehicle, and also upon the overall size of your fleet purchase, the design phase may be very short and basic, or even virtually non-existent. However, if yours is a large expensive fleet purchase and the vehicle type is very specialized, suppliers and manufacturers will be much more willing to make additions or modifications to the vehicle in order to meet your specific requirements. In this case, the design phase can be very comprehensive and extend over a number of months or even years.

These decisions regarding which vehicle to purchase — when individual vehicles will undoubtedly have conflicting strengths and weaknesses appearing from the trials — are often not easy. Careful thought has to be given to the organization's higher level priorities with respect to how this new vehicle will fit into the overriding organizational structure and strategic planning.

*In Stage 2: Forming A Project Team and Points Allocation*, we discussed the general allocation of points during your project. Now let's look at some additional, more detailed considerations specific to choosing a final vehicle. The method used to allocate points must be standardized across all of your scoring areas. Suppose one area — for example, Technical Merit and Performance — allocates 100% of the available points to a single vehicle and the other vehicles are allocated 0% of the points. That type of allocation may give one area a disproportionately weighted effect on the final decision of which vehicle to purchase. This is especially the case if all other areas allocate a percentage between 0 and 100 to each potential vehicle. Such an inconsistency risks accidentally skewing the expensive selection process; it also potentially wastes the money the company has spent on the vehicle trials process. In short, the available points in each area should if possible be allocated to each vehicle based upon a scoring matrix used for all the vehicles. The single exception to this rule is if you decide to allocate fixed scores of 100% and 0% for every respective area of evaluation, such as mobility, mechanical, cost, etc. By allocating the points in the same manner for all areas, you will ensure that one area does not have a disproportionately weighted effect on the outcome. However, this approach can still potentially skew the final results, as shown in the following example. As such, I recommend that you do not use this approach unless for essential reasons. (It is discussed here only for completeness and as a warning against its use).

Figure 4-13 provides a summarized example of a completed vehicle purchase final scoring matrix. In order to detail how the scores were achieved during the vehicle evaluations, a full final scoring matrix will also list the subheadings evaluated for each area. For example, the Mechanical Performance heading may include subheadings such as Engine/Powerplant, Wheels and Drive-train, Electrical, and Lighting, as well as the total points awarded in each area. Which subheadings and how many will be included are dependent upon the specific items you chose to examine, test, and score during your trials. For clarity, I have eliminated all sub-headings in this example. If we use the maximum scores below from Stage 2 Table 2-2 as an example, the points shown in Figure 4-13 may be the final scores for three competing vehicles, awarded exactly as per the trials results.

| | | | |
|---|---|---|---|
| **Vehicle Purchase Cost/Price** | **1500 pts** | **=** | **15%** |
| **Mechanical Performance** | **4000 pts** | **=** | **40%** |
| **Mobility Performance** | **2500 pts** | **=** | **25%** |
| **Spare Parts and After Sales Support** | **1200 pts** | **=** | **12%** |
| **Other Areas** | **800 pts** | **=** | **8%** |
| | **10000 pts** | **=** | **100%** |

The example in Figure 4-13 is the way the Final Scoring Matrix should be compiled, where each vehicle is allotted the exact points and percentages gained during the trials. However, I will now explain the potential allotment of points from this matrix into a single 100% scoring matrix (Figure 4-14). This method can be used to generate a less-detailed final output and keep some of the finer details of the trials results in-house, if required.

As you can see in Figure 4-13, Vehicle A received the highest score, 1400, for Vehicle Purchase Cost/Price. Vehicles B and C were awarded 900 and 400 points respectively. If we were to use the single 100% scoring method, as shown in Figure 4-14, Vehicle A would automatically be awarded the maximum score of 1500 points in this area because it came first for Vehicle Purchase Cost/Price. The other vehicles would be awarded 0 points.

For Mechanical Performance, Vehicle C came in first and is awarded 2300 points in Figure 4-13. Using the single 100% scoring method, the score is changed to a maximum score of 4000 points, with the other vehicles being awarded 0 points, as shown in Figure 4-14.

In the area of Mobility Performance, Vehicle B gained the most points under practical evaluation, with a score of 2350 points in Figure 4-13. Therefore, it is awarded the maximum score of 2500 points using the single 100% scoring method. This method is continued for all evaluated areas, as shown in Figure 4-14.

Again, if the single 100% scoring method has to be used, it must be applied to all evaluated ar-

*Figure 4-13, Summarized Final Scoring Matrix*

| Evaluated Area | Vehicle A | Vehicle B | Vehicle C | Remarks |
|---|---|---|---|---|
| Vehicle Purchase Cost/Price | 1400 (From a maximum of 1500 available) | 900 | 400 | |
| Mechanical Performance | 2280 | 1640 | 2300 | |
| Mobility Performance | 2020 | 2350 | 1160 | |
| Spare Parts and After Sales Support | 780 | 700 | 720 | |
| Other | 640 | 0 | 310 | |
| Total Points | 7120 | 5560 | 4890 | |
| **Percentage Score** | **71%** | 56% | 49% | |

*Figure 4-14, Single 100% Final Scoring Matrix*

| Evaluated Area | Vehicle A | Vehicle B | Vehicle C | Remarks |
|---|---|---|---|---|
| Vehicle Purchase Cost/Price | 1500 | 0 | 0 | |
| Mechanical Performance | 0 | 0 | 4000 | |
| Mobility Performance | 0 | 2500 | 0 | |
| Spare Parts and After Sales Support | 1200 | 0 | 0 | |
| Other | 800 | 0 | 0 | |
| Total Points | 3500 | 2500 | 4000 | |
| **Percentage Score** | 35% | 25% | 40% | |

eas to prevent one area from potentially skewing the overall result. With this method, you will not display matrix subheadings — showing the breakdown of points is unnecessary as you only want to show final scores.

As you can clearly see from Figures 4-13 and 4-14, using the single 100% scoring method has the potential to affect the final scoring and vehicle placement if not used carefully. The different results occur despite the original points from the evaluation being the exact same when used to generate both final matrices.

In Figure 4-13, Vehicle A is the leading final contender and consequently the vehicle you would purchase with standard scoring. Based on the points accumulated from the trials and evaluation, Vehicle A finishes first and is the best vehicle overall. Furthermore, the overall scores for Vehicles B and C from the trials, as shown in Figure 4-13 are also quite low — so low you may have doubts regarding their overall suitability for purchase, regardless of Vehicle A's score.

However, if using the single 100% scoring method it can potentially skew the final purchase results and decision towards a particular vehicle and you must be aware of this risk when scores are being allotted at the conclusion of the trials.

In Figure 4-13, Vehicles A and C are scored very closely for Mechanical Performance. Only 20 points separate the two. Still, Vehicle A has considerably stronger scores overall. However, when single 100% scoring is used, as shown in Figure 4-14, the large number of points available for this single area (Mechanical Performance) risks the entire final score being skewed towards purchasing Vehicle C. The danger of allocating too many points to any single area, such as in most business projects Vehicle Purchase Cost/Price, can change and negate your entire trials results. This wastes all of the time and money you have spent on conducting detailed trials and vehicle evaluations. You must be vigilant for such risks.

## Stage 10: Design Phase and Defining Vehicle Baseline Additions

Once the final decision has been made regarding which vehicle is to be purchased by the company, a project enters what is known as the Design Phase. This can be a misleading title for a project where an off-the-shelf vehicle/equipment solution is being purchased with no changes. However, it is so called because this is the point in the purchasing process at which additions or modifications to the vehicle baseline design must be defined, if so required.

The extent of changes to a vehicle baseline during any of the stages of a vehicle's purchase, or indeed throughout its lifetime, are dependent upon a number of things: the original specifications of the vehicles you purchase, the level of preparation you carry out to clearly specify what you require from the purchasing process, the extent of modifications to the proposed vehicle, and how well proven the unmodified vehicle design is in service — both internally within your organization or externally elsewhere in the world.

Production companies for specialized vehicles do not have tens of thousands of pre-made vehicles, perfectly meeting specifications, sitting in factory parking lots like standard family cars, waiting for individual customers to simply choose the model which they like the most. This is largely because of the vagaries of individual customer's requirements in the specialist fields and environments for specialized fleets. However, if you are buying a fleet of relatively standard vehicles, such as light vans or articulated tractor units and trailers, the design phase will be very minor and a lot of the following will not apply.

Essentially, when you walk into your local car dealership, you are largely restricted to the off-the-shelf options offered by the manufacturer. The purpose of each vehicle is tightly defined by societal norms, i.e., most private cars are used for driving primarily on paved surfaces, to and from work and leisure activities at a reasonable speed. Under normal circumstances, your civilian car supplier does not allow you to change the specifications of every item to meet your own individual requirements. For example, you may be allowed a choice of radio types, but you cannot easily specify any radio from the thousands which exist on the open market and have the manufacturer engineer the vehicle's radio mountings around it. You certainly cannot change the fundamental engineering elements of your vehicle and have them rework their manufacturing line to suit. Of course, you can purchase vehicles and have such work completed by your own maintenance staff or by a third party company if required, but that is a different matter.

For a large enough purchase quantity, specialist, or military vehicle, producers often allow you to specify your options first, for which they will then submit a proposal and often modify or reengineer their vehicle if required. This not insignificant fundamental difference from small private vehicle purchases contributes to the high cost of specialist vehicles (as do other qualification requirements to meet military and industry-specific standards such as SAE, STANAG, MIL-STD, MODUK DEF STAN, and national standards). Fleet requirements for specialist vehicles are much more specific to each individual customer. They may include on-road versus off-road usage profile, load carrying capacity, crew carrying capacity, mounted systems types to be fitted, dismountable crew kit and equipment storage, dangerous goods and materials storage, wheeled or tracked, communications suites, and protection levels. The list of potential requirements specific to each customer can be long and varied.

A majority of the changes required during the design phase of a large fleet purchase will have become obvious throughout the vehicle trials. They should be sought in order to rectify areas in which the chosen vehicle did not outperform the non-successful vehicles, to improve areas in which this vehicle did not meet desirable characteristics of the compliance matrix, or to allow the vehicle to meet national legal or role specific requirements.

There is significant risk involved in baseline additions and modifications during the design phase, as previously discussed in chapter 3. As such, it is essential that a full list of baseline changes be defined immediately after the final vehicle selection. Before choosing to finalize the changes being considered, a careful discussion has to be undertaken with the manufacturer, in order to establish any secondary engineering alterations required to the vehicles as a consequence of each individual change or all of them. It is essential that strict change control is exercised at the beginning of and throughout the design phase. The various personnel, operators, mechanics, area specific specialists, and others offering opinions and requests regarding potential modifications can cause a vehicle to be changed beyond what is reasonable for the envisioned role.

The project team must give significant thought to the list of modifications and additions for the chosen vehicle. The list should not have to be further modified as the design phase progresses. In addition, any changes must not cause the safety and effectiveness of the vehicle to be compromised after entry into service with your company.

It is unlikely that there will be secondary effects for any vehicle project with a number of baseline *additions*. However, it is almost certain that there will be significant secondary or tertiary effects to any vehicle project where *modifications* are being implemented. Therefore, it is important to accept that even after a carefully defined list of additions and modifications is established, modifications will undoubtedly cause a number of further changes over coming months, hopefully minor, as the new baseline is tested and produced. We must remain conscious that our accurately estimated timeline for design and production may change during the design phase due to the testing and alteration of each required modification.

Restricting any modifications to those on the original list at the outset of the design phase helps us ensure a minimum of unexpected delays to the project delivery schedule. The primary aim of establishing a new baseline early in the design phase of the project is to avoid a process of continuously making further additions or modifications and leaving the overall project in a constant state of flux. This is known as freezing the scope of work.

Such a situation of on-going changes during the design phase also results in significant inefficiencies for the manufacturer and their suppliers. It can cause issues with an almost exponential increase in time lag for the logistics trail necessary to supply us with spare parts and maintenance equipment for a new vehicle. Spare parts manufacturers cannot ship the necessary spares to the manufacturer or our organization, if the manufacturer has not yet finalized which components will be used on our new vehicle. Such logistical issues will result in on-going delays in the provision of not only the vehicles themselves, but also Special Tools and Test Equipment (STTE), vehicle kits, workshop equipment, spare parts, and maintenance or operator documentation — all of these essential to enable the successful introduction and operation of a new vehicle fleet.

The problems caused by these delays can be

further compounded if, despite delays in the support apparatus for a new fleet, the timeline for vehicle delivery is set in stone to suit contractual delivery dates defined by your company at the outset of the project. Manufacturers that simply ensure the vehicles are delivered by a particular date, regardless of whether the necessary ancillary support has been put in place beforehand, can cause very serious problems. These problems relate not only to the logistical support for the new vehicles, but also the quality control standards being implemented on your vehicles during production. A lack of sufficient quality control in vehicle production may take your company years to resolve or recover from.

**Guaranteeing that a new vehicle fleet arrives by a particular date is much less critical than ensuring it arrives in a suitable condition and with the necessary support in place to function effectively from Day.1**

It is often an aspect of large vehicle purchases, especially those for organizations with their own in-house maintenance, that the delivery of all ancillary support equipment, spare parts, and training of personnel precede the arrival of the purchased product. Be aware that a more inefficient and troublesome process can sometimes be implemented, whereby everything effectively arrives concurrently over a period of weeks or months with the new vehicles. As the customer, it is best to avoid this scenario.

With such a process, delays will often result because items are shipped from the vehicle manufacturer to the customer in a haphazard fashion as they arrive from third-party suppliers. Undoubtedly you will receive the parts you need immediately, at the end of the spare parts supply process — and the parts you do not need for a number of months or years will arrive first!

Wherever possible, the arrival of the new vehicles themselves should be the last delivery in any complete purchasing process that also involves spare parts and ancillary support equipment.

## Stage 11: Pre-Delivery Inspections

Pre-delivery inspections (PDI) are the most important quality control aspect of a vehicle purchase project that will be conducted by the buyer.

If your PDIs are conducted correctly, they provide an opportunity to examine in detail the purchased product before paying the manufacturer. If necessary, PDIs allow the buyer to refuse to accept the new vehicles should the quality of the vehicles be deemed to be below the required standards. Due to the high level of activity during the construction phase of a vehicle project, and the timeline demands that will often be in place, we must be aware, as purchasers, that at various times during fleet production, the highest standards outlined by producers in their business plan may not have been met for every vehicle rolling off the production line.

Therefore, project financing must allow for travel, as required, to the manufacturer's facility during and after the production process. Approval for this travel can often be difficult to obtain when finances are tight. However, without such visits as well as sufficient time and resources being given to the accompanying inspections, you are effectively purchasing a product or fleet of vehicles blind.

*Would you ever design a new house, wait for it to be built, and then buy it on the word of the seller without ever viewing it yourself?* Or buy it if there is a time limit on the viewing that allows you to have only a cursory look at the outside and not see the entire house first? Of course not.

Hence, project financing must allow time on site at the manufacturer's production facility for very detailed, pre-delivery inspections on your new vehicles. Without sufficient budgeting for this time from the outset, there will be a negative effect upon the rollout of the overall project. You could face significant delays before you can start using the new vehicles, or the new vehicles may suffer from negative perceptions about their reliability and maintainability due to a lack of sufficient quality control prior to delivery.

There are several straightforward aspects involved in conducting pre-delivery inspections. Their importance as a tool to minimize post-delivery issues with vehicles cannot be overstated.

**1. PDIs should be conducted at the manufacturer's facility prior to shipping the vehicles.**

Although there are costs involved in sending inspection teams to a manufacturer's facility, they must be viewed in the context of the overall value of any fleet purchase project. Conducting pre-delivery inspections after the vehicles have already been shipped to you — the customer — does not allow the simple and essential ability to refuse vehicles that are substandard. It also does not allow problems to be resolved prior to leaving the factory, where all of the necessary equipment and production supplies are available if required. Depending upon the quantity of vehicles being purchased, your PDIs should involve sending enough personnel (management and mechanical) to the production site to physically inspect a given set of vehicles and also to examine all the production quality control paperwork. A *pre-delivery* inspection is not pre-delivery if it happens after delivery has occurred.

**2. PDIs should have sufficient personnel available in the inspection teams to allow comprehensive examination of all vehicles ready to be shipped from the completed production batch.**

There is no point sending a token team of one or two personnel to conduct a cursory review of vehicles prior to shipping. If there are going to be mechanical issues that will subsequently cause significant problems post-delivery, a cursory examination cannot discern these. For complex specialist vehicles, sufficient numbers of personnel must be sent on the pre-delivery inspection to allow a minimum inspection time of one day per vehicle for two personnel working together — at least for the initial delivery batch. Of course, if you are buying widely available standard commercial vehicles, this section is not applicable.

These inspections must include simple aspects such as checking the security of all hose clips and pipe/hose connections on the vehicle, checking the security of internal and external electrical connections, operation of internal and external lights, operation of all gauges, etc. They should also include more intensive tasks such as stripping all wheel hubs on a random selection of vehicles to check for production errors such as swarf. When defining the list of items you wish to check on the PDIs, your experience from the trials will guide you regarding what areas did not measure up on the vehicles. Also, any areas that were modified during the design phase should be inspected.

The operation of ancillary vehicles' features or equipment — such as tire inflation systems, hydraulic suspension, overpressure filter systems for vehicles operating in hazardous environments, etc — where fitted, should be checked. All vehicles should be road tested and examined afterwards for leaks from any fluid carrying pipes or hoses. This may require a member of your inspection team to ride along with a driver from the company, depending upon the qualifications of your personnel and the driving regulations or laws in the manufacturer's location.

The inspections should check as many individual items as physically possible, on as many vehicles as possible, within the time available. Any quality control related issues will become apparent as the inspections progress. Inspection teams should be managed in order to reallocate time away from items that are proving to have no issues on numerous vehicles. Instead, they should give additional time for inspecting an increased quantity of vehicles for those items that can be seen to have issues arising. The more detailed and comprehensive the pre-delivery inspections are at the manufacturer's facility, the easier the acceptance testing post delivery will be, and the quicker the vehicles can enter service with high reliability.

**3. Pre-delivery inspections must have sufficient time available to them.**

To carry out their job properly, the inspection team must have sufficient time on site without being under such time pressure that they may miss an important issue with the new vehicles. Time must also be available in the overall project delivery schedule, to allow the inspection team to refuse vehicles for shipping if very significant issues are identified. Again, it is effectively pointless to conduct a detailed examination on a fleet worth hundreds of thousands of dollars or more, if it highlights a major issue but there is no higher level support within the company to move back the delivery timeline. A delay may be necessary in order to ensure prior to shipping that the items being delivered are brought up to the expected standard.

One way to avoid the scenario of not having sufficient time to resolve issues prior to shipping is for the delivery timeline to initially factor in a delay of approximately two months, due to anticipated refusal of the first batch of vehicles. This time is dependent upon the quantity of vehicles purchased and the extent of agreed modifications. Should there be no problems upon inspection, it is always preferable to have a timeline brought forward and delivery occur early, rather than having to move it back because of a lack of foresight about production and inspection-related issues. Do not inform the manufacturer about this two-month cushion, nor the reason, as it may be incorrectly interpreted as a lack of confidence in their production practices.

### 4. Pre-delivery inspections require the support and preparation of the manufacturer.

Be certain the inspection team can effectively "hit the ground running" upon arrival at the manufacturer's site. The inspection must be preceded by good communication with the manufacturer to ensure that all the necessary arrangements are in place before your arrival. The manufacturer should be given the total list of tasks that you intend to complete during your inspections. They should also be given copies of the blank inspection sheets you intend to use to record the inspections. Tell them all the tasks you will be carrying out on all vehicles, and tell them that you will be selecting vehicles at random for particular inspection tasks as applicable. The ideal situation is one whereby the manufacturer carries out all tasks, on all vehicles, prior to your arrival. Your inspection team would theoretically then find zero issues and all vehicles can be shipped immediately.

It's important to remember that the aim of pre-delivery inspections is not to catch the manufacturer. The aim is to ensure the effective performance of the vehicles upon delivery and minimize any problems that arise after delivery. An ideal situation is if the manufacturer can make the inspection task easier by ensuring that the vehicles are 100% ready. Obviously the best situation is if the vehicles are ready for inspection after rolling off the production line, requiring no further checks by the manufacturer.

Coordinate with the manufacturer so that all necessary tools are available for your inspection teams to carry out their tasks. Any necessary consumables that may require replacing during stripping and assembly tasks, such as seals or fluids, must be available in sufficient quantity to leave every vehicle ready for shipping upon the inspection team's departure — with no issues beyond those discovered during the inspection which are to be rectified by the manufacturer.

**The aim of pre-delivery inspections is to guarantee the effective performance of the vehicles upon delivery and minimize the amount of problems that arise after delivery.**

If the engine, transmission or driveline / drivechain components have been modified on a vehicle during the design phase, the manufacturer may also be required to run a specified amount of mileage on one or more vehicles prior to or during your inspections. The time required to facilitate this must be addressed both in your project timeline and in the manufacturer's production/delivery timeline.

## Stage 12: Post-Delivery Acceptance

If the pre-delivery inspections process is carried out in a diligent and comprehensive manner, the post-delivery acceptance process should be largely comprised of inspecting vehicles for damage that may have occurred in transit. If you wish, all vehicles can be checked for potential issues that time did not allow for checking the entire fleet during the PDIs. Such a shortage of time would, however, indicate a failing in the PDI process.

During this final acceptance, check areas such as braking force for the vehicles. Have each vehicle complete your national and local environmental or other road standards testing. In military organizations, this testing may not be legally required, but it is still good practice to ensure that the new vehicles have correctly functioning brakes, steering, lighting, etc., prior to accepting and releasing them for use.

All ancillary equipment must be fully integrated and fully functional on the vehicles before final acceptance is issued to the manufacturer. If such equipment is being fitted in the home country by the purchaser, this process should have been pre-planned to commence upon the arrival of the vehicles. The complete testing of ancillary specialized equipment such as radios, lifting equipment, and cutting equipment is necessary before accepting any vehicles —these kit items can place electrical loads onto vehicle power systems, causing problems that may not be identified prior to extensive practical use.

Vehicles should not be released for use upon delivery, until all such testing and acceptance have been carried out to a high standard. If any issues arise, they will only be further complicated by attempting to resolve them while the vehicles are being used on a daily basis. Testing of weapon systems in military organizations should involve firing hundreds or thousands of rounds and not simply a token quantity; a lower amount will not allow continuous usage problems to become obvious. Likewise in the civilian sectors, any cutting, grinding, digging, lifting, or handling equipment must be tested extensively, not simply started up and shut down after a few minutes.

If at all possible, vehicles in the new fleet should not be released one-at-a-time as acceptance is completed. Such a procedure can lead to problems where the end users, receiving the vehicles, proceed with their training and use of the new vehicles, only to result in breakdowns and availability issues which place unnecessary pressures on the acceptance of the remaining vehicles being completed quickly.

A delivery batch of vehicles should either be released for final use in its entirety after acceptance, or it should be accepted and released in an organized fashion of sufficiently sized blocks, such as 10, 25, or 50 vehicles at a time. This process again ties into the question of time. Enough time must be allocated, after the delivery of the vehicles, not only to account for the acceptance testing itself, but also to allow for the resolution of any issues that arise after arrival. This necessity must be communicated early to upper management in the purchasing organization, to avoid the belief that vehicles will be operating the day after they are delivered.

Obviously, it is also essential that personnel managing the post-delivery acceptance process streamline it and minimize any delays. Get the vehicles through as efficiently as possible so that they can begin serving the purpose for which they were bought. As noted earlier, if the PDIs were detailed and comprehensive, then acceptance testing should not take an inordinate amount of time.

## Stage 13: Modifications, Upgrades, and Life Extensions

Many non-technical managers find it difficult to understand that buying a fleet of new specialized vehicles is not the same as buying a family car. The process does not end when the vehicle leaves the factory or is delivered. Pre- and post-delivery inspections and acceptance processes, post-delivery warranty alterations, in-service modifications, mid-life upgrades, and life extension programs are all integral elements on the road to an operationally effective and cost-effective platform.

The manner in which the vehicle trials, selection, and purchase were carried out up until this point will determine the scale of modifications required to a vehicle immediately after it begins to be used. In an ideal world, there should be a minimum quantity of post-delivery modifications made to vehicles. However, in many cases, the more modifications that were applied to your vehicles during the design phase, the more new elements there are now that can fail or have issues after delivery. This problem explains why modifications can extend the timeline for vehicle delivery so much — in order to avoid such issues after the vehicles have arrived, modifications must be comprehensively tested by the manufacturer prior to shipping. Such testing requires time.

**Warranty Modifications**

The first types of post-delivery modification are warranty modifications. In most cases, they result from manufacturing or production issues, omissions, or errors. Warranty modifications will normally be initiated as a direct result of component failure and come under the umbrella of warranty repairs or warranty retrofits. If the process for new vehicle purchases as described in this book is followed — including the measures to prevent drifting into R&D, limiting the scale of modifications to a trials vehicle platform, and detailed PDIs — then the number of component failures should be minimized on a new vehicle.

This limited failure rate applies only if the platform is a proven one. If the vehicle is unproven in actual use, it is very difficult to determine what problems will arise. By carrying out the purchase process in a comprehensive and calculated manner, the number of warranty modifications that arise will be reduced significantly. However, if significant R&D is carried out in a project, then a time and cost allowance must be factored into the vehicle introduction and post delivery period for warranty modifications that will almost inevitably be required. These modifications are exactly what they say; they are carried out by the vehicle manufacturer under warranty.

It is essential that during the purchasing process you have agreed about the warranty terms with the manufacturer. How long should the warranty period be? In general, the vehicle supplier or manufacturer wants to minimize its length because it has the potential to cost them money. On the other hand, the customer would like as long and comprehensive a warranty as possible. If you can agree to a 3-year warranty on a heavily modified vehicle, you would be doing very well. But with a large number of modifications to a vehicle baseline, you want at least a 2-year warranty (24 months) to ensure most major problems are caught within the warranty time period. Without many modifications, 1 year may be a sufficient warranty.

### In-Service Modifications

After new vehicles begin to be used, the more detailed pros and cons of their design will become clear. As a result, you will receive requests from operating personnel for additions or modifications that will better facilitate their tasks. In particular, military vehicle technology can advance so rapidly that within the timeframe from defining the final baseline to the vehicles being delivered, significant technological improvements can have appeared on the market. Also, previously irrelevant but now significant considerations may have arisen due to the operational experience of other users with your vehicle type.

Choosing to make optional modifications to a vehicle after it has entered service can be referred to under the generic term of in-service modifications. These modifications are generally at your own expense and are not covered by warranty.

The modifications may not necessarily be upgrades; they may result in a more usable or ergonomic platform without changing the performance characteristics of the vehicle. If there are mechanical issues that are not defined as faults, and hence not covered under warranty, they may also be dealt with through in-service modifications. An example is continued failure of a component due to a vehicle being used, perhaps forded, beyond its original intended capacity. Such an occurrence is indicative of the usage requirements of a new vehicle being slightly different than envisioned; it is normal that these can arise. Resolving these issues will involve a two-stage approach.

The first stage is improving training and increasing the awareness of the vehicle operators regarding the capability of the vehicle. This training may remove the need for a modification, saving your company from spending much money on the issue. If this first approach does not decrease the vehicle failure rates and the repair costs, the second stage involves evaluating potential mechanical design changes to the vehicles to resolve the issue.

The initial indication that an in-service modification may be required will often be an occurrence of high volume failures and repair costs that indicate a design weakness in a particular area. Making the decision whether to carry out an upgrade is relatively straightforward, but involves a benefit-cost analysis. A simple method to carry out a basic review follows; for a detailed discussion, including aspects such as market distortion, the discount rate, or distributional considerations, you should refer to a book specifically on this subject, such as *Benefit-Cost Analysis: A Practical Guide* (Anderson & Settle, 1977).

The failure rate of the specific component needs to be examined and, if necessary, extrapolated to give an average number of failures per annum. This number can then be multiplied by the individual replacement costs of that component. Don't forget to include wages to cover the average repair time and the number of technicians involved in each individual repair. If the failure in question results in vehicles having to be transported back to a repair base each time, the costs of wages for the recovery personnel and the fuel costs, over an average or nominal transport distance for such recovery, should also be factored. This gives a total cost to your organization of these failures on an annual basis.

The unit cost to resolve the issue by fitting an upgrade to each vehicle should be sought from the manufacturer or from your own R&D department. Comparing the costs of implementing this upgrade on your fleet of vehicles, to the annual repair costs of the failure in question will result in the length of time required, in years, to make back the financial investment in the upgrade. Table 4-1 provides a sample benefit-cost analysis for a component failure for a theoretical fleet of 100 vehicles.

Table 4-1

Fleet Size (S)
Component Replacement Cost (RC)
Number of Failures per Annum (F)
Total Cost of failures per Annum (TC)
Estimate for Modification Design (M)
Estimated Implementation Cost per Vehicle (CV)
Estimated Implementation Cost for Fleet (IC)
Time to Recoup Expenditure (T)

S = 100
RC = $2600
F = 14
M = $26,000
CV = $1400

TC = (RC x F) = (2600 x 14) = $36,400
IC = (CV x S) = (1400 x 100)= $140,000

Time to Recoup Expenditure =
[(IC + M) / TC] = [($140,000 + $26,000)/$36,400] = 4.56yr

According to Table 4-1, it will take approximately 4.6 years to recoup the direct cost of designing and implementing this modification on the entire fleet. If your vehicles' expected life is 25 years and this modification is carried out early, the cost-benefit is clear. This formula can be expanded to include under TC items such as transport costs for failed vehicles moving to/from the workshops and wages for maintenance personnel each time a failure occurred. The more detail that is added to such a calculation, the higher the accuracy of the appraisal.

Any change to the component replacement cost also has to be evaluated with respect to the new estimated MTBF for the modified or replaced components. If, for instance, the MTBF is expected to double (failures occurring half as often as previously), but the individual component replacement cost will also double due to the modification, then most of the benefit is cancelled out. In such an instance, all the benefit is not lost because the vehicle will still be available for use more often. This availability in itself may be reason enough to carry out the modification. Your maintenance personnel will also spend only half as much time conducting repairs due to this issue over the lifetime of the vehicle. However, the financial benefit of the reallocation of wage costs may not be sufficient to compensate for the initial capital outlay on design and implementation of a modification across a fleet.

In order to establish failure rates (rather than take the suppliers' stated rates as gospel), run a test quantity of the modified component, possibly under an accelerated life cycle program, prior to committing to a fleet-wide rollout. It is worth inquiring whether the manufacturer will share the costs of the initial engineering work and such testing. If they have other customers who may subsequently purchase the modification, there can be longer term financial benefits for them.

Also consider the operational benefits of vehicles no longer becoming unserviceable due to a reduction in the frequency of any failure which you are resolving. There will also be an increase in the time vehicles are available for use. In the search for ultra-reliability, any upgrade to a vehicle that reduces failure rates is desirable. Issues that do not result in a mechanical failure or reduction in serviceability, but which are requested by the vehicle operators due to a perceived need when using the vehicles in the field, can also drive in-service modifications. Evaluating these requirements is much less scientific. However, the results are often more tangible to the vehicle operators than maintenance-driven modifications. A decision whether to modify the vehicles often rests on the effectiveness of the requesting personnel in convincing an organization's management about the benefits of each individual change.

It takes competent management to understand that issues will always be noted with the daily functioning of a new vehicle while it is finding a place in your organization. These issues may require in-service modifications shortly after entry to service or later in its life. Often, if these modifications are not implemented at an early stage in a project's projected life, the enthusiasm and political will to pursue them can wane, and they can be shelved simply for lack of effort.

Setting modifications aside for a long time before pursuing them often results in a lack of appreciation within organizations of why they are required; the vehicles have been seen to be functioning without these changes for years. There is also a reduction in cost-benefit as the remaining vehicle lifespan decreases. The key argument against modifying a new vehicle is often simply that it is a new vehicle. When early modifications are not implemented, potentially great vehicles can remain simply good or average, impacting on turnover and hence profit. The argument against modifying an older vehicle is often that it has been operating successfully, so why make changes. This can mean an aging vehicle becoming progressively less effective and less competitive in the operational environment; your organization's equipment does not keep pace with competitor's improved technology and, hence, your competitors will have reduced costs and higher levels of efficiency.

In-service modifications need to be viewed beyond the context of immediate requests and needs. You must also consider how any modifications may change the vehicle's performance in other areas and with respect to future roles or mission tasks. It is usually a wise expenditure to facilitate vehicle additions or modifications requested immediately after entry into service. The cost of such modifications are often low, both with respect to the project cost to date and the remaining length of service of the new vehicles. Hence, they can provide significant cost-effective benefits over the life of the vehicles.

**In order to ensure the best performing product and best value for the significant sums already spent, in-service modifications that are requested due to increasing experience with a new vehicle, should be facilitated early wherever possible.**

It is important that a percentage of the overall purchase project budget be offset for in-service modifications that may be requested to vehicles after delivery. This percentage is included within the project value allocated to ancillary requirements (up to 40%) and needs to be set aside early in the purchase program for a new vehicle.

**Upgrades and Life Extensions**

Vehicles are no different than other machines in that they have a design life. They also do not have a definite end of life. The nature of vehicles is that their overall design life, is a function of the design life of tens or hundreds of separate components, each with its own design life and each integrated together in the total platform. These components scale up into sub-assemblies, assemblies, and systems. By replacing these different elements of the total vehicle platform with more modern alternatives, we can often significantly extend the total useful life of the overall vehicle.

However, there is always a point at which the fundamental aspects of your particular vehicle type have advanced to the extent that upgrades are no longer cost effective, and a new vehicle is required. You can only establish this point on a case-by-case basis, by carrying out RFIs and comparing the current market technology against the costs and advantages of upgrading your current fleet. An RFI may be as necessary and useful prior to initiating a Life Extension Program, as it was prior to the original vehicle purchase.

Mid-life upgrades and life-extension programs should be an integral aspect of every organization's culture. The technology available in the modern marketplace moves rapidly; significant cost-benefit gains can be made by upgrading a fleet to modern standards within a 10–15 year window after purchase. As alluded to before, a significant danger related to life extension programs is leaving their initiation for too long.

At times there can be a perception that another year will result in better upgrades or more "bang for your buck." Although this can be true early on in the vehicle's life, it can also result in one more year becoming many more years. By the time an organization finds the will to upgrade the vehicles, technology has moved so far forward that a cost-benefit analysis may find greater long-term benefit in purchasing a new fleet. This option may not be financially viable, however, and it thus becomes impossible either to upgrade or replace vehicles —upgrades are no longer cost effective and you cannot afford to initiate a replacement schedule in sufficient time before the vehicles reach the end of their effective life.

Designating a single finite point in a vehicle's

lifespan at which a life extension program should be conducted would be misleading. Generally a life extension program for a particular platform should be reviewed approaching any time towards 50% of the estimated lifespan and onwards. Depending upon the advancement of equipment and technology specific to the role of the vehicle in question, it may be viable to conduct a life extension program earlier. Sometimes changes in the operational environment or a new discovery can lead to a huge technological jump in a short period of time. As a consequence, we may have to upgrade our vehicles earlier than anticipated in order to remain cost effective and competitive.

When conducting a life extension program earlier in a vehicle's lifespan, realize that you may not know what technology is coming; you may wish to conduct another program later in the vehicle's life. When work is conducted prior to 50% of the vehicle's life having passed, it will in many cases be more correctly considered an upgrade than a life extension, and any extension to the platform life can be viewed as a secondary benefit.

Any work being conducted after 50% of the vehicle life has expired should be conducted prior to entering the final third of the expected total lifespan. On a vehicle with a 25–30 year expected life, it is unlikely in most cases that the fundamental vehicle design will lend itself to a cost-effective life extension program after 20 years of technology changes. After such a significant period of time, a replacement program will likely yield the most cost effective result.

The exception to this guideline is with a very specialist piece of equipment where openly available technology improvements do not regularly come to market. We can see this situation in the case of military vehicles such as Main Battle Tanks (MBTs) where, despite huge technological leaps in areas such as weaponry, sighting, target acquisition, and electronics, the vehicle fundamentals have not changed much in a number of decades. Thus, a delayed mechanical life extension program may be slightly more viable, but in many cases still inadvisable.

Evaluating options for a life extension program involves a mix of the two stages previously described for in-service modifications. Tangible savings from resolving failure issues and improving MTBF should be considered alongside the benefits of improving the operational capabilities on a more practical level. An essential feature of any life extension program — in order to slow the obsolescence of a fleet and reduce the LTTA for spare parts — is wholesale replacement of obsolete components on your vehicles. Replacement of obsolete components across an entire fleet at one time, instead of one-by-one as they fail, can give a number of benefits. It can reduce the individual replacement cost, reduce failure rates, reduce spare parts lead times and generate maintenance savings. Replacement of Non-COTS, Non-MOTS, or even MOTS parts with COTS, if available, can also be advantageous.

A life extension program is unlikely to be carried out more than once. Therefore, proper consideration and time must be allocated to evaluating every area of the vehicle's capabilities with respect to lessons learned since its introduction. The life extension program should be planned so that you minimize the future operating costs and gain the maximum extension to the vehicles life for the cost of the program.

Consider the protection levels of military vehicles when looking at a life-extension program. The benefits of increasing protection levels are clear, especially in a world where the threat of ever ingenious and effective counter-vehicle IEDs (Improvised Explosive Device) and similar devices is constantly growing. Armor upgrades are an easy item to justify to an organization's paymasters, either alone or as a big ticket item that other upgrades can then be tagged onto. However, with present technology, increasing protection levels often comes with significantly increased weight. Also, do not forget that certain classes of military vehicles are designed for specific environments and specific threat levels. There comes a point where if greater protection is required, it becomes an operational matter to employ a heavier class of vehicle ranging upwards from motorcycles to MBTs, and all vehicles in between.

A vehicle is usually purchased with the maximum protection level available in its class for that time period. Although protection for new vehicles keeps improving, it is unlikely that within 15 years the protection will have become significantly lighter or that it will be suitable for your older vehicles without major vehicle modifications beyond the cost-effective scope of a life extension program. Any secondary effects to the vehicle's mechanical properties and load carrying capacity must also be considered carefully before adding on more weight, either heavy armor in the military, or larger capacity role-specific equipment in civilian industries.

When we consider vehicle age versus cost of maintenance, it is widely held that as equipment ages, the maintenance costs and failure rates increase. This perspective often contributes to justifying a life extension program. However, it would be remiss of me if I failed to note that a number of studies do not support this assertion. Several factors indicating why analyzed data may not reflect this intuitive decrease in performance due to age are given in *The Effects of Equipment Age on Spare Parts Costs: A Study of M1 Tanks* (2005, RAND Corporation).

This book suggested that analyses of age-related costs which do not show a cost increase, often fail in their accounting for increased labor costs with respect to repairing more frequent age-related failures. This is because labor costs often come from a separate budget than spare parts procurement; in some cases, labor costs are not accurately recorded at all due to data limitations. Failure to properly record task-specific labor costs is unfortunately a common feature of many modern maintenance facilities. External contract costs are also not often recorded in enough detail to reflect an increase in expense due to age-related failures. Because maintenance personnel numbers do not increase as the fleet age increases, the labor costs appear to be fixed.

There are a variety of other reasons that the effects of vehicle age on your maintenance costs can fail to show up accurately. A lack of detailed digitized maintenance history often does not allow you to identify the true age of a vehicle, which will undoubtedly have undergone significant component replacements during its lifetime. Spare parts procurement budgets are often not adjusted to account for equipment age or to trend upwards; therefore, they appear as if costs are remaining static in line with the static budget. At lower levels, maintenance needs that are not met often are not being tracked or recorded; maintenance personnel often work around problems. Some studies show that the majority of age-related failures occur on less expensive vehicle components. These issues can fail to carry a significant enough individual cost to highlight the age problem when each failure occurs.

Budget shortfalls are often viewed in a holistic manner, without accounting for each element of the shortfall. Instead, they are met through an intervention such as a single large influx of funding from another budget or source to meet the shortfall at the end of the fiscal year. If budgets remain static and the requirement for vehicle availability and readiness does not reduce in line with increased age-related failures, unofficial procedures and personnel going outside the standard supply systems to get the necessary parts can explain how the increased demand appears to be met.

Examples include in-house manufacturing, maintenance-to-maintenance transactions, attaching a component purchase onto an official repair bill, or locally purchasing components. Cannibalization can also arise in order to meet parts shortfalls; it does not show up in the accounts unless it is outside of work hours, and your organization pays overtime.

These costs are often age related or highlight failures in the maintenance system; they can help obscure any shortfall in the maintenance budgets attributable to vehicle age. There are many reasons that upgrades and life extensions are cost effective, but it is often difficult to highlight these reasons when age-related costs are hidden or obfuscated on a daily basis.

It is also possible that the maintenance structures in a large organization may strategically delay replenishment of spares for a particular vehicle type or percentage of the fleet until budget pressures ease, for example, during the crossover of the fiscal year. This delay, however, not only carries the problem of insufficient budgets into the end of the next fiscal year, but it also introduces a lack of spare parts for a significant portion of each fiscal year leading up to the end of year parts purchase. This lack of parts can cause an increase in the percentage of vehicles that are unserviceable late in the year, and reduces the capacity of the maintenance system to address any unforeseen failures. It also generates maintenance backlog and increases maintenance pressures at the beginning of the fiscal year while the system tries to catch-up on outstanding maintenance tasks.

These issues all contribute to the difficulty of accurately analyzing age-related costs on vehicle maintenance.

In *The Effects of Equipment Age on Spare Parts Costs: A Study of M1 Tanks,* the authors attempted to account for all of the above mitigating issues when evaluating the effect of age on spare parts' costs. Their results nonetheless showed no positive age effect on actual expenditure or on spare parts' costs. However, the authors note, as in other studies:

> *These results...should not be taken as evidence that age does not affect maintenance costs, only that these data do not show such a relationship...actual expenditures should not be used to estimate a cost versus age relationship...a complete analysis of the effects of age on maintenance costs is not possible without additional data, particularly labor data.*
> *(RAND Corporation, 2005)*

This leaves us with a clear and rather ironic situation regarding the evaluation of vehicle life extension programs. While reviewing any proposed program to achieve a life extension for our vehicles, we can continue to consider possible maintenance savings that may result, remaining aware that we may not be in a position to prove such savings with concrete data later. When making the case for a life extension program, it is advisable to place the majority of the focus on any anticipated improvement in spares availability and reduction in lead times, improved operational effectiveness of the platform, and the expected availability/reliability benefits. Reduced maintenance times are also a possible benefit that has a secondary financial benefit. Anticipated reductions in maintenance times should be evaluated, based upon the manufacturer's Mean Time To Failure (MTTF) of the upgraded replacement components versus the manufacturer's MTTF of the old components, originally fitted to the vehicles. As we have discussed, it may be difficult to accurately evaluate the actual repair times for a new component without very concise historical labor data.

Unfortunately the difficulty in proving age-related costs limits our ability to include reductions in maintenance costs in any proposal to conduct a life extension program. But this difficulty does not negate the other overall benefits to your organization or ancillary financial benefits to other budgets caused by a life extension program or vehicle upgrades. If your organization does have very detailed records on spare parts usage, labor costs, and the age-related effects discussed in this section, you may be able to prove some age-related costs to help make the case for your specific life extension program.

## Chapter 5

# Ancillary Requirements

The lifespan of new vehicles can be significantly affected by the organization meeting — or failing to meet — ancillary purchase requirements. These are the items requiring expenditure other than on the vehicles themselves, such as support and maintenance equipment or vehicle housing. These ancillary items allow vehicles to be operated effectively for the purpose they were intended — from the moment they arrive and throughout their lifespan.

Up to 40% of the total financing available for a fleet purchase project may need to be allocated to ancillary requirements. This focus and expenditure on ancillary requirements applies regardless of whether the vehicles are unique, specialized, or simply additions to vehicles already in service. It also applies, in particular, in the case of most military specific vehicles; however, the exact percentage will depend upon what ancillary facilities you already have in place prior to initiating this specific vehicle purchase program. The importance of ancillary requirements cannot be overstated, yet it is often difficult to communicate or justify upwards through an organization's management structure. The reason for such a potentially large amount not being spent on the product itself is simply to ensure that all new vehicles are actually available for use soon after delivery, and that they give long-term value for money.

There is little benefit to an organization in receiving delivery of a fleet of shiny new vehicles if there are no maintenance and repair arrangements in place, or if the facilities for the storage, preparation, and management of these expensive assets are not created. Unfortunately, gaining support for such ancillary expenditure from higher management can often be difficult. A combination of factors come into play: first, the simplicity of the reasons this expenditure is required, second, ancillary items may have to be purchased form a different budget than the new vehicles, and third,the contradiction with the experience of purchasing a simple private vehicle such as a family car, which requires no immediate investment in support mechanisms. In the latter case, no ancillary expenditure is required because the manufacturer or dealer usually has support facilities and ancillary mechanisms already in place, and the garage at the family home for storing the vehicle already exists.

For all their capabilities and performance, modern vehicles remain essentially machines composed of numerous smaller mechanical sub-assemblies. These sub-assemblies involve no witchcraft or mysteries. They each operate in a particular manner and each operates best under a particular set of circumstances, be it the chemical reactions in a catalytic converter, the movement of clutches internally within an automatic transmission, or the performance of the vehicle's finish or paint coating. If ancillary requirements are not facilitated sufficiently at the initial purchase stage, then vehicles will not operate as intended or will suffer increased failure rates and vehicle downtime.

This chapter looks at the largest ancillary purchase requirements likely to require consideration with respect to your fleet purchase. These should be put in place, or already exist, prior to any new or additional vehicles being delivered to you. Having the necessary additional resources available, such as additional maintenance staff or equipment in place prior to the vehicles' delivery, also provides a secondary benefit of increasing work turnover leading up to the delivery date. This increase will allow you to clear maintenance backlog and free up capacity to deal with the additional maintenance burden of acceptance and post-delivery testing when the new vehicles arrive.

**Maintenance, Spares, and Support Facilities**

The facilities required to successfully maintain a fleet of vehicles are a direct function of the usage profile for those vehicles. The level of care given to the vehicles by the operators and the quantities of vehicles in service also have a direct impact.

For example, one of the most common methods of defining the usage profile is simply between on-road and off-road vehicle usage. The vibrations and strains of off-road use cause a significantly higher amount of wear on a vehicle over its lifetime than paved roads.

Likewise, a lack of care or attention by vehicle operators can result in smaller maintenance issues not being addressed prior to their devolving into more significant problems. Having a higher number of vehicles in service can also result in less strain on the remaining units. A lower number of vehicles in service can result in almost 24-hour operation of individual units; consequently, a very intensive maintenance schedule is required to keep all operating efficiently. Sufficient quantities of vehicles in service will allow vehicles to be withdrawn from service to undergo minor maintenance tasks before any mechanical problems degrade.

The maintenance facilities available must also be physically sufficient for both the vehicles currently in service, plus the additional burden of new vehicles coming on-line. They must be sufficient in both size and turnover capacity. At risk of stating the obvious, if your maintenance facilities have simply performed adequately for a number of years with your current fleet, an increase in the size of that fleet will require a corresponding investment in your maintenance capacity. Without such investment, the maintenance turnover respective to the size of your new larger fleet will be reduced, with a consequent detrimental effect on the condition of your vehicles. Of course, if your new vehicles are simply replacing your present vehicles on a one-for-one basis, then the size and turnover requirements are unlikely to change significantly.

If there are an insufficient number of vehicles in service to meet your operational requirements while also allowing a sufficient maintenance turnover, their deteriorating condition will require an increase in maintenance capacity. You can probably see the link between fleet size and maintenance turnover capacity forming here.

This ancillary requirement for maintenance capacity includes the number of maintenance personnel available, the physical space available in the maintenance facilities, and the availability of workshop and personal tools.

The location of maintenance and repair facilities, respective to the home base of the vehicles, is also important when considering lifetime costs. Depending upon their condition, transporting or driving vehicles to a central repair facility costs money, both directly and in the form of non-operational hours spent. With the ever rising price of fuel, this cost (in addition to the costs of extra off-road time) must be weighed against the price of devolving more of your maintenance to local centers. If your organization already has maintenance facilities in a number of locations, then the costs of increasing their tasks to include third-line maintenance or higher may be minimal in comparison to the transport and related costs of using centralized maintenance. If, however, your organization does not have a network of maintenance facilities, then the start-up costs may be prohibitive. Evaluating matters such as this is essential to ensure that you get the best and most cost-effective use from your new vehicles when they are in service.

Any new vehicle purchase must have a sufficient availability of spare parts to deal with the initial breakdowns after the vehicles enter service. The importance of this requirement cannot be overstated; it increases in importance if the vehicles are of a type not already in your fleet, and even more if they are heavily modified or have new design features.

For example, if you carry out in-house maintenance and the vehicles are of a type you don't operate yet, then a sufficient purchase or supply of frontloaded spare parts will be required. This purchase is needed in order to ensure that the vehicles remain available for use when technical faults occur after delivery. If your maintenance is carried out by a third party under contract, you will need to ensure that your maintenance contractor is sufficiently aware of the new requirements, and they have sufficient initial parts quantities in stock. Note I state when and not if technical faults occur; faults are inevitable and will occur at some point. It is important that these parts match the most likely faults that will appear on the vehicles in the first few thousand km of use.

Any recommended initial supply of spare parts received from a manufacturer must be checked by your own maintenance personnel, whether they are in-house or contracted. It is quite possible a manufacturer will list initial spare parts that an experienced eye can quickly dismiss as not required in the short term. This may not be intentional; the list may actually be computer generated based upon the manufacturer's profile algorithms for vehicle usage. Examples of components that may be front loaded in larger quantities than required include parts that are not maintenance intensive and should not require regular replacement, such as axles and chassis brackets. By all means, you can receive a nominal quantity of such components to deal with unforeseen circumstances or failures. However, the recommended spare parts lists supplied by the manufacturer must be checked to ensure the recommended spares quantities are realistic.

Ensuring an initial availability of spare parts is an essential requirement, as the following experience demonstrates on the adjacent page.

The key lesson to take away from the example on the following page is that the ancillary requirements, such as spares availability, must be assured prior to vehicle delivery. Not every circumstance can be foreseen, but if we fully prepare for the most likely circumstances — or, as in this case, the most likely failures — then the unforeseen and less likely occurrences will have less of an impact on the initial operation of our new vehicles.

In early 2007 I assisted with the delivery of a new vehicle fleet for a close friend working in an Irish civilian logistics company. Spare parts had been front loaded as recommended. In fact, two years' worth of front-loaded parts were delivered as their maintenance was about 60% in-house. In this instance, the vehicles themselves were worth almost $80,000 per unit. The types and quantities of front-loaded parts were recommended by the vehicle supplier using projected failure rates. These failure rates were established from a complex usage profile matrix, based on the vehicles' projected service with the company. The required parts were mostly delivered on time, so initially all was going to plan.

Unfortunately, as is often the case in complex business projects, unforeseen circumstances tend to arise at the most inconvenient time! You can imagine our surprise when shortly after delivery, within a few hundred km of use, vehicles began failing due to faulty differentials. As it transpired, the problem originated with the original equipment manufacturer (OEM). Representative of the best "Monday morning syndrome" on the production line, the rear differential on a limited number of vehicles had not been washed out after machining; they were full of swarf. When vehicles started being used, this swarf damaged the differential seals.

As a consequence, one by one the vehicles went unserviceable and had to be taken off the road. The seals were worth approximately $6.60 each and there were three in each differential. Although such an issue is not ideal, it is not by itself a major problem. However, none of the specific seals were included by the manufacturer of these REEFER vehicles in the initial deliveries for the front-loaded spares supply because they were not expected to fail so soon. To make matters worse, they could not be sourced locally at that time and the vehicle distributor had none immediately available.

The result was that nine tractor and trailer units with a total value of nearly $720,000 were unserviceable for weeks due to $19.80 worth of seals on each vehicle or a total value of $178.20. This single issue caused not only huge financial damage to the business, but also had a major impact on the usage of the remainder of the fleet. Although the situation was challenging, the front loading of spare parts thankfully allowed most other failures on the remaining newly delivered vehicles to be repaired as quickly as possible, thus helping to mitigate the impact of this fault to some extent.

**Vehicle Housing**

The long-term benefits of housing vehicles indoors when not in use are well known. Avoiding the corrosive effects of weathering on vehicles over a period of decades can not only extend the life of a vehicle, depending upon climate, but also reduce maintenance costs. This effect is due in part to reduced deterioration of the vehicles' components from environmental factors, such as expansion and contraction due to cold storage temperatures and warm operating temperatures. It is important that housing is not only available, but designed to facilitate these reduced costs.

> *A simple example of the environment causing premature failure would be two air conditioners: one in a hot environment and the other in a cool one. Which would you expect to fail first: the air conditioner in a garage in California or the one in a garage in Finland? The odds are in favour of the one in California failing first, the one in the hotter environment.* (Borris, 2006)

For vehicles using modern batteries, such as Gel or Absorbed Glass Matt (AGM) type batteries, it is important to address simple considerations such as electricity sockets being available in vehicle parking areas, to allow trickle charging — smart charging — when vehicles are not in use. Modern batteries can be very effective at providing power in difficult conditions such as temperature extremes, but the trade-off is that they are often more sensitive to damage from not being cycled regularly.

With the cost of each modern specialist battery often stretching towards or over $1000, the value of trickle charge facilities for a large fleet is clear. In the case of battery-powered vehicles, their battery banks can cost thousands of dollars.

If you wish your new vehicles to operate over a period of two or more decades, then a facility to keep vehicles dry is also important for the condition of more expensive commercial machinery or military armored vehicles. Placing vehicles into short- or long-term storage in a damp climate, while wet, can have an accumulative deteriorating effect, especially upon sensitive electronics and bodywork. The integrity of armor protection on civilian or military protective vehicles can also be compromised by such factors.

Hence, vehicle housing should be equipped with effective under floor heating or a means of maintaining the internal environment at a suitable temperature and humidity level to allow drying of vehicles after extensive use.

**Communications and Other Equipment**

The price of modern communications or surveillance equipment or, in the military, the price of a weapon system can add a significant cost to the total value of a vehicle. It is important that these ancillary items and other necessary equipment for the practical operation of the vehicles in their envisioned role — such as NVE (Night Vision Equipment) and driving aids — are allotted a sufficient quantity of the total purchase budget at the outset.

It may be difficult to gain the necessary support for such expenditure after vehicles are already in service. The variety of such ancillary equipment is virtually unlimited, and will be determined by the purpose of your new vehicles and their role in your organization.

# PART TWO
# MAINTENANCE SYSTEM

# Chapter 6

## Vehicle Reliability

Reliability: *"Able to be trusted; predictable or dependable"*
*(Collins English Dictionary, 2009)*

*"Capable of being relied on; Dependable."*
*(The Free Dictionary.com, 2012)*

Modern civilian logistics can involve the road transportation of loads over very significant or long distances across the United States, Europe, Siberia, Africa, and other large land masses. In these instances, the timely delivery of a product and the efficiency of the logistics operation overall is dependent upon the reliability of the articulated trucks and similar vehicles involved in the physical movements of goods. Each single breakdown that occurs has the potential to involve repair arrangements with associated companies in foreign countries, where your logistics routes are operating or passing through. These breakdowns can impose a hefty direct and indirect cost burden onto your organization. Significant complications to the organizational delivery system can be caused by breakdowns if there is a major fault with a vehicle, or if there are no alternative vehicles available in the area at short notice.

In a military context, the modern military in most Western nations is expected to be agile and maneuverable, operating over large geographical areas in a dispersed manner. This expectation is often the case in missions undertaken in accordance with Chapter 6 or Chapter 7 of the United Nations Charter. These can occur in large and sparsely populated countries such as those on the African or Asian continents. Scattered tribal villages spread over large areas of responsibility, each being hundreds of kilometers apart with little or no indigenous capability to provide maintenance support, place significant demands upon the reliability of vehicles and equipment.

It is clear, therefore, that the nature of modern military and civilian operations requires fleets of vehicles to be ever more self-sufficient.

In *Repairing The Panzers*, Lucus Friedli describes the movement of groups of Panzers in World War II and the strain caused on the machines of that era:

> *A marching plan also needed to include stops and breaks to regroup the formation and to check, maintain and supply the vehicles.....the Pz.Kpfw. Tiger needed a first stop after 5km, then one stop every 10-15km.*
> *(Friedli, 2011)*

Although modern vehicle technology has thankfully moved on from a requirement to halt individual vehicles or a convoy every 10 km, vehicles in many fleets, perhaps most fleets, remain more unreliable than necessary. Implementing proper procedures and practices to improve the reliability of our vehicles would significantly help to reduce the quantity of failures and, consequently, the quantity of unplanned maintenance.

Ever tighter profit margins, and the pressures of rising fuel costs cutting into those remaining profit margins, are amplifying the effects of each individual breakdown on the bottom line of companies operating fleets of vehicles. As operating costs increase and profits decrease, there is less margin remaining with which to maneuver budgets and expenditure to redirect funds and take up slack in different areas. The costs in these areas, such as those caused by breakdowns, can often be reduced or eliminated through the efficient implementation of an effective fleet management and maintenance system that improves our reliability.

At present, military organizations from a number of nations involved in long-range operations or long-range patrolling often travel with a mobile repair truck or mobile workshops in their convoy as a standard operating procedure (SOP). The nature of civilian vehicle operations does not usually lend itself to this method of operating. However, even when these support vehicles are used in the military context, they often lack the maneuverability and terrain-crossing ability of the vehicles they are intended to support. Therefore, they become a limiting factor in otherwise dynamic and maneuver-intensive operations. If we look past this obvious limitation of having these support vehicles in place, we can see that the true limiting factor is the unreliability of the primary vehicles. It is this unreliability that drives the need for constant local support availability.

*Unreliability* is perhaps too strong a term. After all, the mechanical reliability of most modern vehicles has improved in leaps and bounds over their predecessors. Thus, a more accurate appraisal of the situation may be to use the term lack of *ultra-reliability* in our vehicles.

Such a lack of reliability in the vehicles we operate will be indicated by the failure of vehicle components during use. These failures will normally manifest as one of the five commonly recognized types I describe below (and as listed by R.Keith Mobley in *Maintenance Engineering Handbook*) and on the following page.

Critical, hidden, and functional failures will usually manifest as a loss of integrity of the component, requiring immediate replacement or repair. Evident and incipient failures can cause a loss of operating efficiency that may not require immediate replacement, but will at some point in the future in order to return the vehicle to 100% operability.

As modern operational requirements develop in the military sphere and in ever more competitive civilian industries, both nationally and internationally, there is an ever developing need for vehicles to operate more reliably. These vehicles are also required to operate with minimal levels of support. Longevity and extended Mean Time Between Failure (MTBF) or Mean Miles Between Failure (MMBF) have become watchwords of the modern employment of vehicle fleets.

**Critical Failures** — A complete failure of the component causing it to cease operating or be unable to begin operating; for example, a light that fails to turn on.

**Hidden Failures** — A failure that is not obvious to the operator during use of the equipment; for example, a vehicle brake light that is not working.

**Evident Failures** — A failure that causes a change in the performance or output of equipment, thus indicating to the operator that something is wrong; for example, a failed turbocharger that causes the vehicle performance to noticeably degrade.

**Incipient Failures** —Failures for which there is a slow degradation of performance in small unnoticeable increments; for example, a slow puncture in a tire.

**Functional Failures** — A failure that follows on from an incipient failure. When the degradation of performance becomes noticeable to the operator and is no longer acceptable, the incipient failure becomes a functional failure; for example, when a tire is nearly flat and it becomes obvious to the driver through steering wheel vibration or other signs.

**Note that where the term MTBF is used for calculation purposes, it can be substituted with MTTF (Mean Time To Failure) and vice versa. This is because both of these measures are calculated in the same manner, by dividing cumulative time in service by the cumulative number of failures. The technical difference between these two terms is that MTBF refers to the failure rate of repairable items, whereas MTTF refers to the failure rate of non-repairable items.**

It is clear that reliability is essential to our fleet operations. However, when we think about the reliability of individual components in relation to MTBF and MTTF, we should also consider the following;

**Anyone on the street can determine which individual spare part or repair arrangement is the cheapest to buy for a single failure. But we want to determine which gives the best fleet-wide value and return over numerous failures.**

Non-repairable MTTF items on vehicles — such as light bulbs, fuses, bearings, etc. — are usually replaced complete as is. In the case of a MTBF component, when it is repaired, it may or may not have the same reliability value as when it was new. In practice, we see that a repaired component will sometimes not perform to the same reliability standard as a brand new replacement, or it will often not have the same expected life before suffering another failure. This is a very important consideration for your spare parts' replacement system to analyze. When managing their budgets, the spare parts personnel cannot look simply at the direct repair vs. replacement cost of your components. They will need to take the comparative failure rates of new vs. repaired parts into account when calculating how to minimize your ongoing vehicle repair costs.

There is little value in repairing a component for 50% of the new component cost if the repaired component will fail again within 50% of the running hours of the new component. In such circumstances, you're paying less per individual unit and more per month or per year. Likewise, if you can repair a component to an as new condition for 50% of the price of buying a new component outright, and they will both last the same amount of time in use, there is little value in purchasing the new component.

There are other considerations that relate to the choice among new parts, spurious parts, or repaired parts. (A spurious component is interchangeable with the original part fitted by the vehicle manufacturer, but comes from a different component manufacturer. Spurious parts will often be less expensive than original parts, but will not be manufactured to the same quality standards.) Your choice may further depend upon the nature of the component. For peace of mind and security purposes, you may not wish to repair some components such as brake linings and you might prefer to purchase new pads.

**A lower cost repair may not be cost effective if a new replacement would have operated for a proportionately longer time relative to its cost.**

However, as with many aspects of vehicle maintenance, considerations such as the choice of new, spurious or repaired parts, cannot be analyzed nor a decision made without comprehensive records of failure rates. The use of repairs instead of purchasing new replacements, buying and using refurbished parts instead of new, or spurious parts instead of original, can all have significant value for money implications for your organization.

In most cases, the spare parts that we purchase will come with an approximate figure for usable life from the supplier. However, the only way to comprehensively measure spare parts alternatives (new/spurious/repaired) is to use all and compare the purchase cost with respect to the actual lifespan obtained in use with your fleet.

So why are we discussing the failure of individual spare parts so much? Because the overall reliability of your vehicle is a direct function of the reliability of its hundreds of individual parts.

Warranty arrangements on alternative types of spare parts can also have significant cost implications, but again these cannot be accurately evaluated until you have a complete awareness of the failure rates. Suppose a spurious part is purchased for 75% of the price of an original, with a 6-month warranty instead of the original part with a 1-year warranty. Initially the cost-benefit of the original part seems clear. However, this may not be the case. If based on your vehicle mileage rates, there are fewer failures of the original part within the first 12 months of usage, but they all then catastrophically fail around the 14-month mark, the 12-month warranty is of negligible value to you. The spurious parts may, however, all fail within their 6-month warranty; thus, your long-term replacement costs are going to be much less because their replacements are free of charge. This is only a very generalized example to illustrate the point because more frequent failures will also carry ancillary cost implications related to lost on-road time, recovery and maintenance expenses, etc. Such ancillary failure-related costs must also be accounted for when determining which alternative is the best value for money.

Reliability for our vehicles is deeply intertwined with and directly affects the availability of our vehicles. The availability of a vehicle is the exact same at any point in time as the vehicle's probability of not failing at that same moment. The reliability (%) of a vehicle is also directly equal to its availability.

This availability can be calculated historically for a given period of time as follows:

$$\text{Availability (\%)} = \frac{(\text{Total Time Available} - \text{Downtime}) \times 100}{\text{Total Time Available}}$$

*(Borris, 2006)*

As fleet managers, it is important for us to accept that a vehicle will never be 100% reliable. There are always going to be a limited number of unplanned failures. However, the availability — and, hence, reliability of vehicles — can be maximized by implementing a preventive maintenance system with the aims of maximizing availability/reliability and minimizing costs.

Such a system is ultimately a balancing act between different types of maintenance being used within a single organization at the same time. A continuous effort to reduce the quantity of unplanned or reactive maintenance is also required to maximize the effectiveness of our fleet resources.

Note that a preventive maintenance system and a predictive maintenance system are not the same. A predictive system involves scheduled or ongoing monitoring of system status and subsequent replacement of components when it is observed that a failure appears imminent. Monitoring temperature, using electron fractography, oil viscosity or odor monitoring, endoscope or fiberscope monitoring, electrical resistance monitoring, and much more are used by predictive maintenance systems to predict the best replacement time for a component or lubricant due to the observed degradation rate. There are associated costs involved with predictive maintenance regarding time allocation within the inspection schedule. On the other hand, a preventive maintenance system uses analysis of past data on failure rates to establish the best replacement time for components in order to prevent the majority of failures. The associated costs involved with preventive maintenance relate to the on-going recording of failure data to be used later.

**Predictive systems can be said to look forward to determine when to replace components. Preventive systems can be said to look backward to establish the best replacement times in the future.**

In general, the consequences of failures can usually be mitigated by either eliminating/reducing the failure frequency of occurrence—how often they occur—or by taking action to directly minimize the consequences of said failures. Either way, the first priority in resolving a vehicle failure issue is and should always be to determine the root cause of the failure. This can be different to the immediate cause of failure. The process of Root Cause Analysis (RCA) is basically like peeling back an onion and establishing the underlying cause behind the obvious. This is carried out, repeatedly, until there are no further underlying causes identified as leading to an individual failure.

It is important to remember the fundamental point: a vehicle is no different to any other machine. It is, essentially, a collection of cross-functional, independent, and interdependent parts working together. Any machine can suffer failures and any machine can have its failure rate reduced through the application of sound engineering and maintenance principles, thus becoming more reliable.

Before calculating the availability or reliability percentage, as mentioned above, it is important to define what constitutes downtime within your specific organization. It may be a vehicle that has suffered a complete failure and requires recovery and repairs to regain mobility. When discussing specialist vehicles, *downtime* may be any time when the unique equipment fitted to the vehicle is unavailable for use, even though the vehicle itself is fully mobile. An example is a refrigerated truck that is fully mobile, but unavailable for use because the refrigeration unit is unserviceable. As such, the definition of downtime in a multi-type fleet will normally vary from one vehicle type to the next.

The operators of modern vehicles, be they civilian drivers crossing a continent or soldiers on the ground in an operational theatre, have an increasingly low tolerance for equipment failure. They expect ultra-reliability to be the normal standard because that is what they have known from the time they were children growing up in the modern world.

> *...people or organisations acquire assets because they want the asset to do something. Not only that, but they also expect their assets to fulfil the intended functions to an acceptable standard of performance...for any asset both to do what its users want and to allow for deterioration, the initial capability of the asset must exceed the desired performance. Thereafter, as long as the capability of the asset continues to exceed the desired standard of performance, the user will be satisfied. On the other hand, if for any reason the asset is unable to do what the user wants, the user will consider it to have failed.*
>
> *(Moubray, 2007)*

From the moment new vehicles roll off the production line, infrequent failures are key to gaining and maintaining operator confidence. When it comes to operators of new, unfamiliar vehicles, perception seriously affects reality. If operators have lost confidence in their equipment, the effect on your operations cannot be overstated. Once this confidence is lost, it is very difficult to regain. Although this is particularly evident in the military, it is also very applicable to the commercial environment. Vehicle drivers departing a location, concerning themselves with the likelihood of a vehicle mechanical fault occurring, are not concentrating on the task at hand. In the worst case, a perception of unreliability can lead to non-use of expensive equipment assets because management leave vehicles sitting idle.

Such an expectation by personnel of ultra-reliability is not unfounded. Consider, for example, the modern household television or refrigerator. This equipment can operate for decades without suffering a catastrophic failure. Modern radios found in almost every household can be used for hours every day over a period of years, without a single failure.

Another example comes from Richard W. Price (*Army Logistician,* 1999), who gives the example of the Single-Channel Ground and Airborne Radio System (SINCGARS). He stated that at the time of his example in 1999, SINCGARS had achieved a reliability of 3,000–3,500 hrs MTBF. Thus, reliability increased two-to-threefold over the original requirement of 1,250 hrs MTBF. At the time of writing this book, according to one of SINCGARS European sales agents, SINCGARS reliability has now increased to a MTBF of 5000 hrs, a fourfold increase over the initial requirement of 1,250 hrs, which the radio originally failed to meet! This very clearly highlights the potential increases in reliability that are possible over time through both engineering changes and technological improvements.

Richard Price also discusses the MMBF (Mean Miles Between Failure) of the M1 Abrams Main Battle Tank. This vehicle achieved 304 MMBF for Block 1 vehicles and, despite increasing complexity for the Block 2 M1A2 design, the MMBF was increased through engineering to 419, an improvement close to 40%.

As you can see, reliability as the normal state for equipment is a reasonable expectation for modern operators.

When we consider reliability in the context of our overall fleet, there are implications related to value for money and cost effectiveness. It is, therefore, important to note that a fleet reliability of 66%, for example, can be viewed as equivalent to effectively having one-third less vehicles actually serving in our fleet. In a large fleet, this can be tens or hundreds of individual units unavailable at any given point in time. This equates to a large amount of the original purchase capital, possibly tens of millions of dollars or more, sitting idle over the lifetime of the vehicles. Even if a high average daily reliability can be achieved — for example, 90% in a fleet of 100 vehicles, each worth $200,000 — the unavailable 10% is a total of $2,000,000 of assets effectively wasted each and every year over a 20–30 year lifetime, or $60,000,000 over a theoretical total 30-year lifetime of the fleet. Thus, if we can increase the reliability of our fleet by even a small 5%, we will save $30 million over the lifetime of the vehicles. Fleet availability improvements achieved at less cost than this amount can be said to be cost effective. Obviously, the scale of such cost-benefit depends upon the type of depreciation you use on your fleet and the method you use to determine the usable life of each vehicle, and how far into the lifecycle your vehicles are at the time of improvements.

# Chapter 7

# Implementing Reliability Improvements

The ideal method to achieve ultra-reliability is always for manufacturers to design for reliability from the outset of a new vehicle program. A very simple example — failures caused by corrosion or rust — can be eliminated at the design phase through the judicious use of sealing products.

**In many cases, an intentional design failure can be incorporated into an expensive component, whereby a sacrificial element is included to prevent subsequent failure of a more expensive component. Sacrificial elements are rarely acceptable in road vehicle design because they can be used as a way to avoid having to resolve engineering flaws. However, in some other areas of transportation they are essential, such as the use of sacrificial anodes on the propeller shafts of seagoing vessels. The important components in such instances are protected from degradation by electrolysis causing current to more readily flow away from sacrificial zinc anodes, degrading the anode instead of the propeller shaft, and thus ensuring the reliability of the critical component.**

Over the past few decades, companies striving to improve their products have also found that designing for reliability is much less expensive than redesigning products to improve reliability at a later date.

Contrary to popular perception, the Tiger tank in World War II was not significantly less reliable than its predecessors. However, a lack of designing for reliability was demonstrated in certain areas of its engineering. For example, asphalt roads tended to rapidly wear out the inner road wheels. Of itself, this issue may have been difficult to reduce and was bad enough alone, but it was even further compounded. The design of the vehicle was such that an inner wheel change was also a very difficult task requiring at least 1 days' work for an experienced crew. Therefore, the rapid failure of this part on asphalt roads demonstrated a lack of design for reliability, which was compounded by a lack of design for maintenance regarding the replacement time for this component.

For vehicles already in service, however, there is little alternative but to improve reliability on the existing platforms through a program of in-service modifications and upgrades. By focusing on the most commonly failing components or assemblies on a vehicle, and replacing or modifying these first, followed by the next most commonly failing items and so on, the reliability of the overall platform can be incrementally improved. Nothing in life is perfect, however, and this system of improvements has its drawbacks. It is often prohibitively expensive; in my experience, it can also be agonizingly slow. It can also lack the corporate level support that would be present for a defined vehicle upgrade project that has a clear beginning, end, and cost. On a large fleet of vehicles, if there are common failures of a relatively minor part, the costs of redesigning the component and fitting the modification to every vehicle can quickly add up.

A decision will have to be made as to whether modifications for improving reliability, or indeed for other reasons, will be carried out:

1. Completely in-house
2. Supplied by the vehicle manufacturer, but fitted in-house
3. Supplied and fitted by the vehicle manufacturer

Each option has advantages and drawbacks regarding cost, time, and—significantly—responsibility for subsequent unforeseen failures on the altered design. Sometimes the level of expertise or research budget in your company will restrict the options you can pursue to 2 and 3 and your preferred method of proceeding may not be the ideal choice.

Limitations in either the financing available for modifications, or the facilities available, have led a number of modern military organizations to operate a home fleet of training vehicles and a second operational fleet of the same platforms. The second fleet have been fully kitted out with the various expensive extras and modifications necessary in theatre, whereas the home fleet just have the basic equipment. Avoiding the requirement to fully equip all vehicles with the most expensive modern ancillaries keeps the costs of kitting-out or upgrading the vehicle fleet to a minimum. Although this can reduce costs, it is important to ensure that operators are not being trained on vehicles that are fundamentally different in aspects such as weight or operator positions. Dependent upon the scale of differences between operational and non-operational vehicles, there is something even more significant for our operations — vehicle maintainers trained on a home fleet and sent to different locations worldwide may not be sufficiently familiar with all the systems on the operational fleet — so as to ensure the most effective and speedy maintenance in theatre.

Operational planning delays and problems can also result when there are too many differences between variants of the same vehicle platform. Lower specification vehicles at home cannot easily replace those abroad which have been damaged or lost operationally.

Military mission profiles are expected to change for many nations over the coming decades. Therefore, it is essential that fleet managers and organizations develop methods to improve the overall reliability of their vehicles. Defense industries have to strive to develop ultra-reliable vehicles and products. Such developments are needed to remain competitive and to meet their customers' increasing demands for vehicles to have smaller logistics support requirements during operations. Unfortunately, as is often the case with such matters, one of the biggest obstacles to new procedures and practices can be an organizational resistance to change. In many cases, vehicle manufacturers have a vested interest in not completely eliminating failures —a significant portion of their sales is comprised of spare parts and after-sales support.

The road to ultra-reliability on any vehicle begins with the component manufacturer furthest back in the supply chain, through the engineering departments of vehicle manufacturers, and on to the end users and maintenance managers who must strive to pursue reliability enhancing modifications to the vehicles they already operate.

The introduction of a reliability-focused and driven maintenance system can ultimately stabilize vehicle failure rates and downtime. Vehicles are required to do what they need to do when they need to do it. The only way to achieve this goal through vehicle maintenance is by resolving potential issues prior to them becoming failures.

> *No physical asset operates flawlessly forever. In most organisations, breakdowns are the norm. Quality and productivity losses are high...Since the majority of these deficiencies are manifest as equipment-related problems, for example, breakdowns or maintenance-related corrective actions,maintenance is too often blamed for all problems...In truth the reasons for these inherent problems are shared by all functional groups.*
> *(Mobley, 2008)*

Failures of vehicles are often seen by both the vehicle operators and the management of an organization as mechanical problems with solely mechanical resolutions. More often, however, improving future reliability can involve identifying other root causes of failure. Such causes may be misuse of vehicles or non-implementation of proper procedures by the operators or end-users.

> *Nearly three quarters of all accidents are due to the action (or inaction) of human beings.*
>
> *(Narayan, 2012)*

Any investigation into a recurring fault must begin by reviewing the circumstances prior to a vehicle being declared unserviceable or off-road. It must take account of the operating context, that is, the physical environment and the demands made upon the vehicle by the operators. The overall usage profile of the vehicle must also be taken into consideration.

In medium to large fleet sizes, it is not unusual for some vehicles to be used in a role for which they were never originally designed. This will have a subsequent direct effect upon their condition and maintenance state. Vehicle failures due to misuse, inappropriate tasking, or lack of care by operators, are forms of forced deterioration of the vehicle — as opposed to natural deterioration. These types of failures should be avoided because resolving them is an inefficient use of resources. They are also a waste of the original precious purchase capital while the vehicles are not operating. In addition, they draw on a disproportionate chunk of the maintenance budget because they require reactive maintenance post failure. In some cases, it may be more cost effective to simply sell off vehicles which are being regularly used for an unintended purpose and use the money generated to facilitate the purchase of a smaller quantity of fit-for-purpose vehicles. Despite the lower number of new fit-for-purpose vehicles bought using the proceeds of the sell-off, these will often ultimately perform the required tasks more efficiently. They will also do so while reducing the draw on the maintenance budget.

In many organizations, the majority of maintenance work is wholly reactive, i.e., repairing vehicles as they fail. In many cases, this is because vehicle analytics are not being used; accurate records are not being kept on the causes of either unique failures or repeatable failures, and also on when such failures occur. Without this information, it is exceedingly difficult to identify a starting point for good planned preventive maintenance.

Hence we reach a conundrum of failure analysis; In order to have a theoretically perfect maintenance system, we must analyze failures to identify where to improve. But in a perfect maintenance system, we will not have sufficient failures to provide the required data to pursue any comprehensive improvements. In most organizations, a perfectly functioning maintenance system causing a lack of failure data is certainly not the issue they have to worry about!

There are many proponents of data collection on failure rates through the direct testing of relevant vehicle components. If this type of testing is being used, it must not be carried out solely in an engineering lab environment. I have experienced such testing of vehicle components in the engineering test environment, where the same components that passed bench testing by a significant margin, and were apparently ready for use, then suffered rapid disintegration while put through practical evaluation when fitted to a vehicle. This was despite apparently successful bench testing and a recommendation from a manufacturer to proceed with full fleet-wide installation.

**The only accurate way to determine component failure rates on your fleet is to record failures as they occur in actual usage of the vehicles.**

This recording of failures as they occur in operation of your vehicles allows additional operational factors, such as indirect loadings, vibration, and driver induced wear, to be accounted for in the recorded data.

By implementing a maintenance system driven by prevention and reliability improvements, rather than reaction, it is possible to reduce the occurrence of more significant failures that can ultimately prove more expensive and time consuming to resolve.

> *...not enough attention was paid to maintenance...This situation led to the Panzers being in action for so long that they were either totally lost or small mechanical problems left unattended degraded to where major repairs were needed in the Werkstatt and they fell out for a longer period.... sufficient examples have occurred as to how detrimental this can be on the vehicles.*
>
> *(Friedli, 2011)*

A fully functioning preventive maintenance system can also significantly reduce the quantities of spare parts holdings, as it becomes possible to predict when parts will be necessary. Being able to accurately plan when to replace components on your vehicles allows you to order spare parts only as they are required. When fully implemented, preventive maintenance can lead, over time, towards a state of maintenance excellence, reducing overall failure rates, reducing total life costs, and ultimately increasing vehicle lifespan. An increase in vehicle reliability will also improve the safety of the personnel using the vehicles. Promoting a greater awareness within your organization of the importance of preventing vehicle failures imbues a sense of ownership among the personnel. Your personnel need to realize that care of equipment and performing routine inspections is a vital part of improving availability. Inspections carried out by vehicle operators, and more comprehensive inspections by mechanics, must have the aim of identifying imminent failures on vehicles, and not simply seeking faults that have already occurred.

Preventive maintenance systems require the replacement of components prior to the completion of their useful life. There are direct costs involved in implementing such a replacement system. However, the increased costs of higher replacement frequency are warranted due to the savings in the opportunity costs of lost on-road time, which would be otherwise caused by unplanned maintenance. There are also opportunity cost savings caused by reduced secondary failures throughout the life of the vehicle. Savings are achieved by avoiding these secondary failures and the more expensive damage they can cause to a vehicle. The overall aim is to reach a level where the combined costs of preventive and unscheduled maintenance are minimized. The program can then be adjusted and improved on an on-going basis. Generally a division of approximately 75/25 between scheduled and unscheduled maintenance is a good initial division of maintenance types, which you can then build upon.

The overall preventive maintenance program can be broken down into different levels, based upon the frequency of inspection of the vehicles and replacement of parts. The exact interval between maintenance levels, the number of levels, and scope of work at each level is dependent upon your specific vehicle types and goals.

Generally, there will be four levels of maintenance in a preventive maintenance system and up to five levels in a reactive or predictive system, as shown in Figure 7-1. These are called Lines and range from 1st Line (minor tasks) to 5th Line (very specialized tasks).

Components that can potentially involve 5th line maintenance are often LRUs (Line Replaceable Units). The repair skills for this level of maintenance do not usually exist in-house, except within the largest of organizations. The number of 5th line maintenance tasks in a preventive system are negligible because the failure types that generate 5th line repairs — stripping and repair of a components internal electronic circuits, etc. — are usually pre-

*Figure 7-1 Levels of Maintenance*

| Maintenance Line | Description | Usual Maintenance System |
|---|---|---|
| 1st | Frequent minor work such as checking tire inflation levels. | Predictive<br>Preventive<br>Reactive |
| 2nd | Slightly more significant service items, such as oil or filter changes. | Predictive<br>Preventive<br>Reactive |
| 3rd | Replacement of brakes, test lane inspections, driveline adjustments, etc. | Predictive<br>Preventive<br>Reactive |
| 4th | Overhaul or rebuilding of major components, such as engines or transmissions. | Predictive<br>Preventive<br>Reactive |
| 5th | More detailed and intricate repairs of component failures such as failed ECU circuits. | Predictive<br>Reactive |

vented by the replacement of the entire component as a 3rd line task.

It is very important to consider failure distribution when analyzing the failure rates of specific components. Failure distribution is one of the key elements of preventive maintenance; it will ultimately determine the mileage interval at which you replace parts.

What is failure distribution? It is the frequency of failures occurring at different levels of mileage because failures do not all occur at the exact same mileage interval. The same type of failure on any specific component type will fluctuate on either side of the average mileage at which this failure type occurs. That is, one specific type of failure might occur on average every 10,000 km, but a number of instances of this failure will occur at a higher mileage and some at a lower mileage. This distribution must be taken into account when planning the usage of parts, for both preventive and reactive maintenance requirements. It is illustrated by the following quotation:

> *...if we recorded failures of 100 tires, and their combined operational life was three million km, what can we learn from the mean value of 30,000 km of average operational life? In practice, it is likely that there were very few failures within the first 5000 km or so, and that a significant number of tires failed after 30,000 km. Hence, the actual distribution of failures is important...*
>
> *(Narayan, 2012)*

Unfortunately in many organizations, preventive maintenance does not encompass the entire maintenance philosophy. It often never extends beyond routine maintenance tasks such as servicing, which adds very little toward reducing the number of significant failures.

A comprehensive maintenance system must begin prior to the issue of components and spare parts to workshop personnel. It begins from goods inwards. Components requiring the application of sealants or protective grease during preventive maintenance tasks, such as through the use of the *Lanotec*™ brand of protective products or many others, can be color coded or tagged by the spare parts stores when they are initially received from the suppliers.

This is an adaptation of the Japanese principles of Fuguai, or F-Tagging contaminated areas on equipment for attention. When the parts are issued and fitted to vehicles, the maintainers apply sealant X to parts that are color coded or tagged red, sealant Y to parts that are tagged green, etc. The simplicity of systems such as this can reduce on-going maintenance times and increase maintenance simplicity. Unless there is an agreement with your suppliers to deliver the components already premarked appropriately, such systems do require more investment in personnel within your logistic supply systems, in order to prepare and mark the components as required. However, when these components are issued from the spare parts store and fitted to vehicles, these types of systems can avoid subsequent oversights in the incorrect application of maintenance caused by human error.

In *Maintenance Engineering Handbook*, R. Keith Mobley describes the responsibilities of reliability engineering as follows. Although the language is specific to manufacturing equipment, the principles are the same for all machines, be they vehicles or static production machines.

- *Ensure maintainability of new installations* **(Vehicles)**.
- *Identify and correct chronic and costly equipment* **(Vehicle)** *problems, eliminate repetitive failure.*
- *Technical advice to maintenance and partners.*
- *Design and monitor an effective and economically justified preventive or predictive maintenance program.*
- *Proper operation and care of equipment* **(Vehicles)**.
- *Comprehensive lubrication program.*
- *Inspection, adjustments, parts, replacements, overhauls, and so on for selected equipment.*
- *Vibration and other predictive analyses.*
- *Protection from environment.*
- *Maintain and analyze equipment* **(Vehicle)** *data and historical records to predict maintenance needs.*

In many organizations, the role of reliability engineering and meeting the above responsibilities falls squarely upon the shoulders of maintenance management. In many organizations, human nature can also often lead maintenance managers to generate figures in such a manner as to reflect upon their output in the most favorable light. However, if the principles covered throughout this book are implemented, solid and tangible improvements can be achieved with accurate figures being generated for maintenance rates, which will be confirmed by definite budgetary savings.

Finding the time to carry out all the necessary tasks for a complete review of a maintenance system, to the necessary level of detail, can be exceedingly difficult in the modern high pressure environment of increased output requirements with decreasing resources. However, if a concerted effort is made to improve the maintenance system and move towards a significant proportion of preventive maintenance, there are tangible long-term benefits to the organization, in both operational and financial terms, along with an improved through-life condition and better availability/reliability of the fleet.

# Chapter 8

## Efficiency

When a given output requires less input in time, energy, or raw materials, it can be viewed as being achieved more efficiently.

Although we talk a lot about efficiency in business these days, what we are really discussing is the monetary gains that can be made due to the reductions in time wasted. These monetary gains can be significant.

Efficiency is a much maligned word. This is not because bringing about tangible efficiency improvements is impossible or exceedingly complicated. In fact, these benefits are sometimes not difficult to achieve. However, in the majority of cases, achieving efficiency improvements requires a substantial investment. Not just financial investment, but also investment of effort and time. A determined effort is required initially to evaluate where the biggest gains can be made; this evaluation must then be followed, as necessary, by the expenditure of precious funding to achieve those gains.

However, improving efficiency is often viewed as a means to improve outputs and results without financial investment. In most cases, this goal is simply not realistic. In modern fleets, free improvements alone are often neither practical nor viable if you wish to make significant efficiency gains. In most cases, financial investment is a prerequisite to the greatest efficiency improvements.

Incremental efficiency improvements can be achieved by supplying better or more suitable hand-tools and personal equipment, or by improving organization in the workshop. The provision of tools that improve efficiency, such as impact wrenches, can have a cost-benefit return for the maintenance operation as they reduce turnaround time.

> *Frustrating circumstances in maintenance such as a lack of proper tools, delays and unreasonable pressures, can all combine to cloud judgment and emotion.* *(Reason & Hobbs, 2003)*

Of course, certain efficiency gains can be made in the overall maintenance system through non-financial methods. Increasing the output of maintenance staff by improving their motivation is one such example. An increase such as this will often be driven by loyalty to a maintenance manager and are often short-term increases, which are difficult to maintain. They can also be driven by financial incentives, although financial incentives are a clear cost in themselves. Note that constant pressure on staff simply to work faster can lead to errors and raise safety issues.

Efficiency is fundamentally about the use of time, where small savings each day can accumulate to become enormous time and, hence, financial savings in the course of a year. For example, let's take a workshop with ten maintenance teams, each team consisting of two mechanics. If the individual mechanics each increase their hands-on tool time (HOTT) by ten minutes per day, a total saving is achieved of 16.5 hours per 5 day week, 66 hours per month or nearly 800 hours (equivalent to 100 x 8 hour days) extra maintenance being conducted per year.

> *In the days before the steam engine, we used human or animal power to carry out work. The steam engine brought additional machine power, enabling one person to do the work that previously required several people. As a result each workers output rose dramatically. The value of a worker's contribution, as measured by the number of items or widgets produced per hour, grew significantly.*
>
> *(Narayan, 2012)*

Initially small increments of time can be saved by ensuring the ready availability of tools and equipment, spare parts, accessible vehicle parking, etc. The methods to save a few minutes each day in an average workshop are as varied as they are numerous.

Reducing the time taken up by indirect or support tasks (tasks that do not directly contribute to

> **I once worked with a company that had an employee carpark situated approximately 200 m from the business entrance. There were approximately 100 employees walking between the building and carpark a minimum of twice per day (morning and evening). Taking a conservative estimate — not accounting for delays and personal interactions between the carpark and business — a total of 2 min per person was being lost each time an employee had to walk to or from the carpark. This amount potentially totalled 400 min per day, 33 hrs per week or 1716 hrs per year in lost work time. At an average of $24 per hour, the loss was over $40,000 per year. By relocating the employee parking area to immediately outside the entrance (simply by rearranging some fencing and signage), this time loss was significantly reduced.**

the end goal) and maximizing the effectiveness of the time spent on direct tasks (tasks that directly progress towards the aim of turning over vehicles) is the key to improving efficiency. In the case of maintenance, indirect tasks are all those either before or after hands-on maintenance that occur when a vehicle is not available for use.

Some indirect tasks are as follows:

**Recovering vehicles to a maintenance facility**

The physical location of your maintenance facilities, respective to the vehicle operating locations, is key to improving the efficiency of your vehicle recovery tasks. In civilian organizations, you might establish a network of associated companies in your operating zones with a like-for-like, on-call repair service and facility, whereby you repair an associated or subsidiary company's vehicles at your locations, in exchange for them repairing your vehicles in their country or state. Such agreements are, however, unlikely amongst direct competitors. Contracting companies for on-call emergency repair and maintenance is also used in many cases. Organizations such as the AA in the UK or AAA in the United States are examples of this type of service, although there are many others that specialize in commercial vehicles.

In Chapter 7, I listed the five potential levels of maintenance. In a large organization, the maintenance infrastructure can be spread out over a number of different locations using what's called a 3-Level maintenance system. Every location or workshop will usually conduct Level 1 and 2 maintenance tasks. This can be known as 1st and 2nd line maintenance. 3rd line tasks and higher, however, will often be centralized on a single site. 4th and 5th line tasks may be on a single site or subcontracted off site to a third party.

Usually all 3rd line repairs (and higher) are conducted at a single location. Due to the less frequent occurrence of the more significant types of vehicle failures, it is almost impossible to maintain the skills currency of a widely disbursed maintenance staff for all such tasks. Conducting all 3rd line repairs at a single location also frees up the 2nd line maintenance staff for all of the less significant local repair tasks. Staff in most locations can therefore carry out a higher number of Level 1 and 2 repairs. In addition, centralizing all 3rd — or higher — line maintenance tasks, ensures that there is no requirement to purchase the equipment for higher level maintenance at every workshop location. Higher level maintenance management based at the main workshops, are also better able to maintain an awareness of the occurrence of more significant vehicle faults on the total fleet.

To summarize, in a 3-Level maintenance system, the vehicle operators conduct Level 1 maintenance tasks. At numerous workshops the majority of an organizations maintainers will conduct Level 2 or 2nd line tasks. A limited number of personnel in a single centralized location will conduct the level 3 tasks. In military organizations, the reduction of recovery and repair times is largely dependent upon the chosen maintenance system. Suffice it to say here that in a 3-Level system, inefficiencies due to recovery and movement time are unavoidable. The availability of a sufficiently sized recovery capacity for your total fleet is important. The quantity and locations of recovery vehicles is also important. Sufficient recovery services which will avoid a long wait for recovery when a vehicle breaks down, or avoid delays in vehicles being picked up and transported to the nearest maintenance locations are also important.

The recovery drawbacks of a 3-Level maintenance system can be rectified to some extent by switching to a 2-Level system — discussed in Chapter 12, *Types of Maintenance Systems.* In a 2-Level system, you invest the required resources to enable every maintenance location to conduct 1st, 2nd, and 3rd line maintenance tasks. However, such a switch may involve substantial investment.

**Fault finding on the vehicle**

Time spent awaiting review of vehicle problems and for faults to be diagnosed are also inefficiencies that are necessary, but which delay direct repair tasks. In Chapter 12, *Types of Maintenance Systems,* we discuss the inefficiency of maintenance backlog. Fault finding and diagnosing a vehicle's problems can eat up precious time and contribute to backlogs. The maintainers' familiarity with a vehicle platform can reduce this time, but this is dependent upon the maintainers becoming, at best, specialized in a particular vehicle platform or, at least, familiarized with the particular platforms they are often working on in any multi-type fleet. With limited numbers of maintainers or a limited quantity of a particular vehicle type in service, familiarity with every vehicle type can be challenging to achieve, because the individuals do not get sufficient time repairing each vehicle type.

Richard Price gives the example of vehicles being designed for maximum use of modular replacement. His suggestion leads us to see modular replacement as an alternative or extension to Line Replaceable Units (LRUs). Modular systems taken to an extreme may one day allow assembly of vehicles in a Lego™-like fashion. Vehicle diagnostics, which are used by operators at 1st line, will simply indicate the module number to be replaced. The unserviceable module can be replaced and returned to the second 2nd line location, where it is diagnosed in detail and repaired while the vehicle continues operating with its new slot-in or plug-and-play replacement module. Such a system may seem too good to be true. It would require a sea-change in the philosophy of vehicle design and manufacture by commercial manufacturers and the industrial defense complex, but it would hugely reduce the turnaround time of vehicles during use. On a simpler level, the availability of sufficient quantities of fault finding equipment, such as diagnostic units, can also save time.

**Spare parts supply**

Spare parts supply inefficiencies are perhaps one of the most significant consumers of time in any maintenance system. The inefficient supply of spare parts can also be incredibly frustrating to the staff on the maintenance side of an operation, from the individual mechanic to management level. The supply of spare parts suffers from the contradiction that the raison d'être for these parts is to be available for maintenance. Yet they often have a high financial value and must be accounted for and issued carefully, which can contradict the need to simply ensure they are issued as quickly as possible.

Further complicating the spare parts situation, in some cases, is the erroneous belief by accountants that a spare parts system can be measured and reviewed solely according to the monetary value of the stock holdings. This often manifests itself in an attempt to reduce stock holdings in order to make the maintenance account look less expensive on paper.

It is true that there are costs associated with holding spare parts. These include storage space, heating, rent, building depreciation and maintenance, materials handling vehicles, office equipment, fuel usage, purchase tax for buildings and equipment, insurance, wages of stores personnel, depreciation and degradation of spares in storage, and stores management. These costs are well covered in a variety of logistics literature, but they alone do not paint the full picture.

There are correct methods by which to reduce spares holdings, such as switching to a preventive maintenance philosophy, reducing overall fleet size, or retiring a particular vehicle type. However, attempts to reduce spares holdings are not always due to a comprehensive evaluation of failure rates, demand profiles, and ancillary costs. They are often driven simply by a wish to reduce the total paper value of the stock holdings, which inevitably occurs at the expense of maintenance timings and fleet availability rates. The attempts to reduce stock holdings also often fail to account for the original capital expenditure in the setup of the spares establishment, which cannot be recouped by a reduction in total inventory. Unless a present stores building and staff are going to be fully eliminated by a reduction in spares holdings, then reducing the inventory of in-service parts, which results in a building not being used to maximum capacity, is actually increasing inefficiencies in the system. It is also increasing the potential delays in the delivery of parts to the vehicle maintainers, which are no longer being held as part of the in-house inventory.

In addition, it can sometimes be more economical to hold larger quantities of spare parts, which can be purchased cheaper at the expense of a longer lead time, than to buy the same parts from a more expensive short LTTA supplier. An organization's accounting department will often not realize these details. They will also not realize that every spares purchase transaction an organization has with a vendor costs money in the form of administrative costs. Simply purchasing the cheapest spare parts is not sufficient in order to compensate for these costs. A sufficient quantity of spare parts must also be purchased in each transaction. Purchasing larger quantities requires more storage capacity, but compensates for these other costs.

Good cost-saving decisions, in any aspect of vehicle maintenance, requires detailed recording and analysis of relevant data, analytics, and information related to the maintenance system.

Facilities need to be used to maximum capacity to gain maximum value for money; idle space in a spares store is wasted capital expenditure. This said, insufficient space can result in a higher rate of written off spares prior to usage and increased obsolescence. Thus, a middle ground between too much space and too little must be achieved.

In a military organization, for strategic reasons, there is often the requirement to hold larger quantities of spare parts than in an equivalently sized civilian fleet. In addition to the more rough and intensive use of military vehicles, strategic reserves and deployment holdings are required, as are holdings of parts specific to military equipment that cannot be purchased at short notice. This can also occur in civilian vehicles of unique role-specific design for specialized industry such as mining, exploration, or forestry. There must be a combined approach to spares supply and maintenance at all levels within the organizational structure to ensure that the correct quantities of spare parts are being held.

The priorities of personnel directly involved in maintenance often diverge from those of personnel directly involved in spares management. The former prioritize turnover of vehicles in as short a time as possible, whereas the latter can often make it feel like a battle trying to release spare parts in a timely fashion.

Delays of hours or days in the delivery of spares from your stores to your workshop floor can have a dramatic effect on maintenance turnover times over the course of a full year. Similar to the previous example of saving a few minutes for each maintainer per day, what appear to be minor individual delays in spare parts supply can have a dramatic effect on the fleet over extended periods of time. The personnel involved in the spares supply system must effectively be brought into the fold and made aware of their importance regarding the bigger picture of off-road time for the fleet and efficient maintenance.

Unfortunately one of the most inefficient examples of many spares supply systems is the use of people to carry out tasks that would be better served in the modern day by machines. One of the most common examples is the use of human "pickers" to retrieve stock from individual storage bins on shelf units and forward the stock for distribution. Such a system is labor intensive and inefficient. Modern technology allows the use of pick-and-place robots operating at a much higher speed than humans to retrieve goods. Of course, there are disadvantages to the use of pick-and-place systems, primarily the additional space requirements and capital investment cost. However, for larger maintenance operations, these disadvantages are compensated for by the significant improvement in stock retrieval. Another significant consideration of automated systems is the requirement for standardization of item sizes, although there are ways around this by placing an individual staff member into the delivery cycle to remove individual items from larger bins, or incorporating different types of storage methods.

If a maintenance site is co-located with the spare parts store, the entire requisition and delivery process can be semi-automated from the individual mechanic on the workshop floor, in a complete loop, back to the mechanic, as shown in Figure 8-1. Implementing a pick-and-place robot system is expensive and requires a significant amount of floor space in the spare parts stores. As such, it is not viable for small fleet operators and only suitable for operators of large maintenance facilities.

> *...it is interesting to note that the large carriers using computer-controlled systems have lower and generally more uniform inventory values than those companies using Cardex or bin-control systems.* *(Dolce, 2009)*

Such a system can be achieved by each work bay having a terminal allowing the requisitioning of parts by the mechanic. Upon the click of a button, this list is instantaneously forwarded electronically

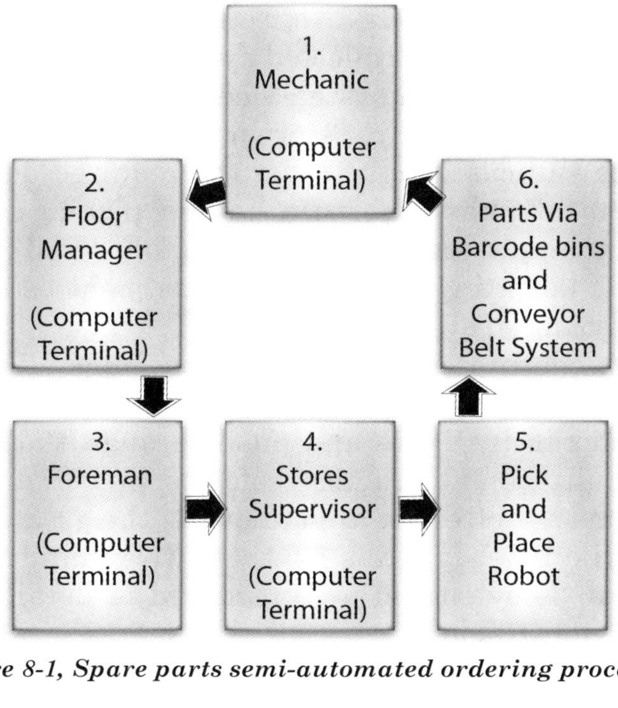


***Figure 8-1, Spare parts semi-automated ordering process.***

to the terminal in the workshop foreman's office for approval; from there it is forwarded electronically to the store manager for issuing approval. The system then activates a pick-and-place robot, which retrieves the parts and transports them directly to the relevant work bay. A system such as this can reduce the spares supply lead times within a large organization to minutes as opposed to days. If a requested part is not in stock, the system adds it to a daily print list of parts for purchase; the lack of immediate availability is instantly flashed up on the screen of the requisitioning mechanic to keep everyone informed.

In the case of different workshops which are not located on the same site, the system picks the parts and deposits them in a collection zone from where they are shipped or collected by conventional means as required. Such a system does require significant investment. The spares store must never be allowed to lack sufficient capacity for the required quantity of parts holdings; otherwise, manual picking of parts from non-automated spill over stores will be required, defeating some of the investment benefit in an automated system.

Modern technology facilitates huge efficiency gains, but imagination and foresight are required in order to instigate and push-through the necessary cultural and practical reforms. Spares delivery systems are often an area allowing the greatest potential efficiency gains for the entire maintenance system. They should be considered in detail when reviewing efficiency.

**Movement of vehicles between repair facilities**

The movement of vehicles between different levels in the maintenance system is discussed earlier in this chapter and in the Chapter 12, *Types of Maintenance Systems* , regarding the 2-Level and 3-Level systems of maintenance. More specific to military systems, vehicles may have to move between different types of repair facilities to undergo maintenance regarding the different and unique systems fitted to an individual platform. These different areas, such as mechanical and communications, can often overlap significantly due to the increased use of electronics in a modern vehicle. There are practical drawbacks in failing to locate maintenance functions within a single site — or preferably a single building — where it is easier for maintainers to discuss issues and offer assistance across spatializations as required. Ensuring that maintainers working in different specializations are based in the same location, allows the ready discussion of technical issues. This avoids your personnel having to communicate between independent and physically separated maintenance sites or workshops. Locating different maintenance specializations can reduce the time to resolve difficult technical problems and help progress the pursuit of maintenance improvements.

**Movement of vehicles back to operators**

The movement of vehicles from a maintenance location back to the operators or end users introduces delays in a similar fashion to the recovery of the vehicle upon initial failure. The reduction of this time is facilitated at a strategic level, through a review of the geographical placement of maintenance facilities with respect to operating locations for the vehicle fleet or its elements. In addition to the inefficiencies of time being lost by having a badly located maintenance structure in your company, there are also costs involved in the form of fuel usage, which can be significant over a period of years or decades. The cost-benefit due to separate maintenance locations again ties into the overall type of maintenance system being used.

# Chapter 9

# Evaluating The Maintenance System

> *The ratio of maintenance cost to total operating expenditure can be quite high, ranging from 10 to 40%. As a result, this item of expenditure attracts a great deal of attention. People do not always recognize the contribution of maintenance in improving short-term and long-term profitability, but are invariably quite aware of its costs. (Narayan, 2012)*

Military vehicles will tend to comprise a larger number of sub-assemblies than their civilian counterparts, again due to their usage profiles and wide variety of required capabilities. It is obviously important to recognize that the amount of time vehicles spend being maintained is time that they are not being used for their intended purpose. Therefore, the overall maintenance system must be set up to get vehicles repaired and back to their operators in the shortest amount of time possible. To achieve this aim, each organization should have its own dedicated maintenance system that best suits its requirements and organizational setup. However, in all organizations, there should be a maintenance management manual available that covers all of the facility procedures, scheduling arrangements, responsibilities, information collection and review systems, and other key aspects of how the organization's maintenance facility operates and interacts with your fleet. It ultimately is a manual for how to run the fleet maintenance facility or facilities. Such a manual prevents any lapse in efficiency should any key individuals drop out of the system for a period of time, due to illness or any other reason. This manual can also be used as a start point when evaluating the entire system, and identifying which areas to improve.

**We need to ask "Why?"** Why are we conducting maintenance? Is it to prevent a future vehicle failure or to resolve a present failure?

**We also need to ask "What?"** What tasks and maintenance should we carry out? Should we complete the minimum necessary tasks to return our vehicles to service, or should we also conduct additional maintenance to reduce the likelihood of recurring faults?

**We also need to ask "When?"** When is the best time to bring a vehicle in for maintenance? The answer is dependent upon the expected timeframe for completing the expected maintenance tasks, because we need to minimize the effects of availability issues on both operations and the state of the overall fleet. We also want to avoid creating unnecessary maintenance backlog.

Although the recommended maintenance schedule from a vehicle manufacturer can serve as a starting point for our maintenance routines, the schedules provided are usually understandably generic (i.e., not customer specific). Such recommended maintenance schedules do not focus on vehicle components and assemblies suffering increased wear specific to your individual organization operating that vehicle. The importance of having the most effective and efficient maintenance system, designed around your own company's unique usage profile and requirements, is equally valid for commercial and military vehicle fleets.

Because vehicle availability can be dramatically affected by the quality of the maintenance being carried out, we must be able to define the overall quality of the maintenance system. Quality can be calculated as follows:

$$\text{Quality (\%)} = \frac{(\text{No. of Vehicles Repaired} - \text{No. of Failed Repairs}) \times 100}{\text{No. of Vehicles Repaired}}$$

This quality percentage can be calculated over a given time period by limiting the value for No. of Vehicles Repaired to your defined time, such as the maintenance quality within the past month, past 6 months, past 2 years, etc.

The number of failed repairs can be defined as vehicles that became unavailable for use, again within a specified time limit, due to a repeat failure of a repaired assembly. A reduction in the quality of the maintenance system can be caused by substandard parts or maintenance error, among other reasons.

For example, if a component is repaired and the MTBF for the repaired component is expected to be 10,000 km, a failed repair could be defined as a failure occurring within less than 10% of the MTBF. Hence, if this component fails again on the same vehicle within 1000 km, it will be considered a failed repair. If an examination indicates that the root cause of this reoccurring failure can be pinpointed to a source different than the original repair, then it may or may not be discounted depending upon your own standardized definition for a failed repair. When establishing your time period for a failed repair of a particular component, it is important not to change the definition until a sufficient period of time has passed to gain an average failure rate. Consistency is required in your recorded data in order to benefit from it. If there are 0% failures appearing after a reasonable monitoring time, then the failed repair time can be increased. If there are 0% failures appearing after a reasonable monitoring time, then the failed repair time can be increased. This monitoring time will be unique to your fleet usage profile. Although as the items on the extremities of the failure frequency of occurrence begin to suffer faults, we would obviously expect to observe failures more quickly in a larger sized fleet.

Highlighting such repeated failures is important and can show mechanical areas of a vehicle platform that require in-service modifications to reduce failure rates and generate future maintenance cost savings. It can also highlight training issues with either your vehicle operators or maintenance staff. There are many other reasons that may be causing such failures to occur, such as low production quality on particular components. But if you are not aware of the repeat failures, you cannot identify the cause or take action to mitigate the damage and costs being incurred. Quality issues could potentially lead to a requirement to find alternative suppliers for the relevant parts.

In addition to calculating the quality of the maintenance system as described, it is also necessary to calculate the Mean Time To Restore (MTTR) your vehicles for each of the major individual faults on that vehicle type. MTTR should be calculated using the total time from when the vehicle went unserviceable up to the time it was again serviceable.

$$\text{MTTR} = \frac{\text{Cumulative Time (hrs) for a No. of Repairs of a Given Failure}}{\text{Cumulative Quantity of Given Failure}}$$

Each vehicle sub-assembly will obviously suffer from wear (direct or indirect) and require preventive or remedial maintenance throughout the lifetime of your vehicles. In many cases, with the ever more widespread integration of vehicle diagnostics across various sub-assemblies, modern vehicles are often no stronger than the weakest link. They tend to self-diagnose and switch into a limp-home or safe mode if a single major sub-assembly is not operating correctly. To avoid expensive and extensive damage to major vehicle systems, such diagnostic linked protective systems are often exceedingly complex on modern vehicles. While saving money on larger repairs, they can in certain cases cause almost as much vehicle downtime on an annual basis as genuine critical, evident, or functional failures. These factors contribute to modern military and commercial vehicles often being maintenance intensive. High maintenance requirements can then be compounded by the intensive and often rough usage profile of vehicles during military operations and also the learning curve of both operators and maintainers of new vehicles.

In order to keep both new and serving vehicles operating at maximum efficiency and effectiveness, the maintenance structures of an organization have to be set up to operate as efficiently as possible. The maintenance structures should be modified to facilitate any increase in fleet size due to each new vehicle purchase, and to facilitate the increase in workload and turnover times due to initial learning curves with new vehicles. Maintenance systems also need to operate efficiently in the best possible way across the entire selection of vehicle types in the total fleet.

The evaluation of a maintenance system must begin prior to vehicle breakdown; it must identify methods to reduce the demand for maintenance through better care and use of vehicles at an operational level. A high standard of basic cleanliness of vehicles both internally and externally allows minor faults to be identified and rectified early, before they devolve into more serious issues.

> *Fundamental to being able to achieve a smaller logistics footprint, of course, is a major reduction in demand for logistics through advances like significantly improved reliability of our equipment and more fuel efficient vehicles.* (Stevenson, 2002)

There are different ways to set up and operate a large maintenance infrastructure, but the overall maintenance system must be driven by the vehicle availability requirements as determined by a particular organization's needs.

The maintenance capacity of an organization will facilitate or restrict the desired on-road availability percentage, but it should not be seen as defining the requirement. This is a common mistake made by non-technical management, who often believe that the maintenance system is a static inflexible function. This approach to vehicle availability is effectively coming at the problem from the wrong direction.

The on-road vehicle availability percentage required within your organization — for example, whether 50% or 90% on-road availability is required — is defined by your organization based first upon its operational and training needs. Only after the required percentage is defined must the maintenance structures be analyzed, designed, and modified to meet the required turnover of vehicles. In this way, we facilitate the necessary on-road percentage. As the necessary percentage increases, the costs involved in its achievement increase. If the total capital investment cost to the maintenance infrastructure is excessive in order to achieve the required availability (or impossible based upon fleet size and failure profiles), we look at balancing maintenance investment with alternatives such as additional vehicles in order to reach our goal.

**The maintenance setup of an organization does not define the on-road state of the fleet. It simply facilitates the required percentage.**

It is important to note that brand new vehicle types in our fleet will take longer to repair while maintenance personnel are being familiarized with the new platform. Vehicles will also be withdrawn for repair more often while operators are familiarizing themselves with the correct operation and the techniques to care for the new vehicles. Building a culture of care for new platforms tends to take place over a period of years rather than months. This factor presents a problem for fleet operation in that it contributes to more downtime for new vehicles. Further compounding this problem are any introductory issues surrounding design reliability and component failures on new vehicles. These factors must be accounted for when defining the maintenance requirements to deal with such vehicles.

Without specific vehicle and fleet-wide details, it is difficult to simply quote a viable on-road state for military fleets of armored vehicles or any other fleet such as commercial tractor units. For new vehicles operating within a fleet that has efficient preventive maintenance, availability can often be up to 98%. I have previously determined in practice that approximately 82% can be viable given specific circumstances. However, this level is wholly dependent, and can vary significantly depending upon usage profiles, operator experience, and vehicle types within the fleet. As such, the level can increase or decrease significantly while still being a realistic and viable percentage unique to your own operations.

The vehicle availability percentage figure is always based upon a logarithmic curve. As the percentage increases, it requires a greater investment in maintenance support to achieve the same gains in addition to those already achieved; as it decreases, more benefit can be gleaned from a smaller investment. The rate of improvement to the vehicle availability percentage is subject to all investment in the maintenance system being used in the most beneficial way possible.

The logic behind this is rather simple. A smaller on-road percentage is usually indicative of more and larger inefficiencies in the maintenance system. As the availability percentage shrinks, the likelihood of basic fundamental problems causing such inefficiencies increases. Although they may be expensive to resolve, fundamental large inefficiencies such as insufficient workshop space, numbers of staff, and quantities of specialist equipment are often relatively straightforward to do so. Proportionate to the percentage increase in vehicle turnover to be expected from these improvements, and hence the improvement in vehicle availability, the investment in these initial improvements will give a significant return if spent wisely.

Such significant returns from any maintenance investment then become more difficult to achieve. As the biggest and most fundamental maintenance issues become resolved, they may generate individual gains in availability exceeding 10%. However, as the overall availability percentage rises above the 85–90% mark, the remaining inefficiencies or problems, which have a smaller scope and are more difficult to address (e.g., hands-on tool time or HOTT issues, spare parts delivery from the

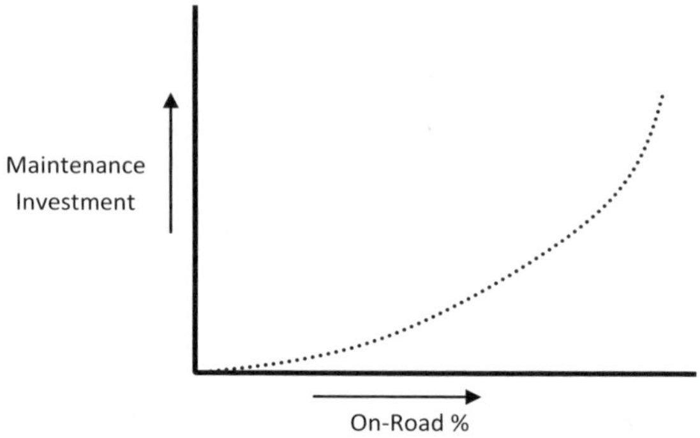


***Figure 9-1, Effect of maintenance investment on vehicle availability.***

stores to the maintainers), may be able to generate gains of only a few percentage points each. These latter remaining inefficiencies are also often the ones that were left until last to address; the costs of quantifying them and then resolving them are often higher proportionately to said smaller gains. This point is illustrated in Figure 9-1.

By way of an example to illustrate the point; if a maintenance system is generating a vehicle availability of 60%, we may determine that the physical space capacity is lacking. By investing to expand the workshops, we can hugely increase the maintenance turnover and the vehicle availability percentage may jump to 80% for, say, $800,000–each 1% cost us $40,000. In order to address the remaining less obvious inefficiencies, we will need to look at more complex issues such as HOTT. To effectively monitor HOTT and address this issue, we may invest in clock cards and workstation clocks for our staff at a cost of $150,000. But the efficiency gains we can later generate using knowledge from this technology, will perhaps only give us an increase in vehicle availability of 3%, which will be achieved over a longer period of time, and it will require more investment in time to operate the system going forward. Each 1% cost us $50,000 plus the ancillary costs over the longer implementation period. From 83%, we may identify a lack of workshop hoists and decide to add an additional one for $70,000; it may increase our turnover — and hence vehicle availability — by only 1%, and so on. As we address the larger, more obvious problems, the remaining problems, which are less obvious, are usually more complicated to identify and address; they are more expensive for the smaller return achieved.

For a static maintenance system infrastructure, as the fleet size increases, the improvement to a fleet's operational state or availability percentage that can be gained from purchasing additional vehicles also decreases relative to the cost of each additional vehicle. This is because the increasing burden on the maintenance infrastructure caused by an increase in vehicles counteracts the benefit of adding such vehicles to the fleet. At a certain point, dependent upon fleet size, maintenance capacity, and the required on-road availability percentage, improving both maintenance turnover and reliability can have a greater effect upon the overall availability of vehicles than by adding more vehicles to an over–stretched system through new purchases.

With the average wage of a mechanic in the United States being about $34,000 in 2012, according to *Salary Wizard Basic Market Pricing Report*, there are clear potential savings to be made by employing additional mechanics or improving maintenance performance, instead of purchasing new vehicles. Such benefits can also be achieved by reallocating a portion of any new vehicle funding to maintenance improvement.

At the opposite end of the scale is a point where the maximum benefit to operational availability is gained from purchases, where improvements to maintenance are no longer as cost effective. In organizations worldwide, there is often a tendency to deal with availability problems by arbitrarily purchasing additional vehicles, without first taking a holistic approach and analyzing if availability can better be improved by distributing a portion of funds to the support systems for the vehicles already in service.

**Ultimately, the more stretched or degraded the maintenance system has become, the more benefit can be gained from direct investment in that system.**

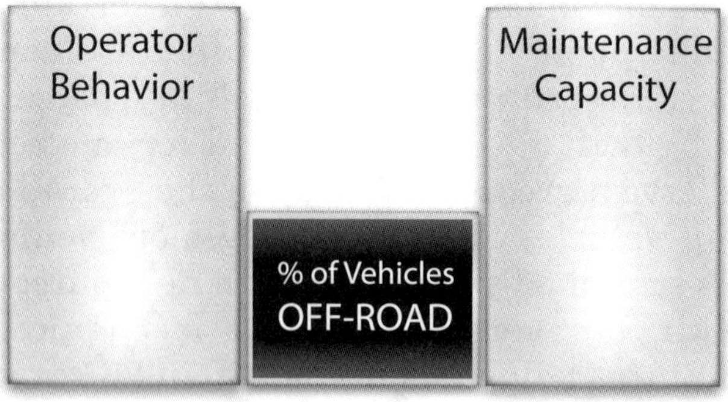


*Figure 9-2, Effect of Operator Behavior and Maintenance Capacity on the percentage of vehicles off road at any given time.*

The on-road state of a vehicle fleet is also generally affected by two key variables: 1) the operator behavior (and experience/familiarity) with the vehicles and 2) the overall size of the maintenance capacity. These variables are demonstrated by Figure 9-2. The importance of these variables is not often fully appreciated and accounted for during the daily operation of fleets. When either of the two outer columns increases or decreases, the central column moves in an opposite direction. If the central column moves, then both outer columns move in tandem, in the direction opposite to the central column.

**If your Maintenance Capacity increases in your fleet, as shown in Figure 9-3, the percentage of vehicles off-road will usually decrease.**

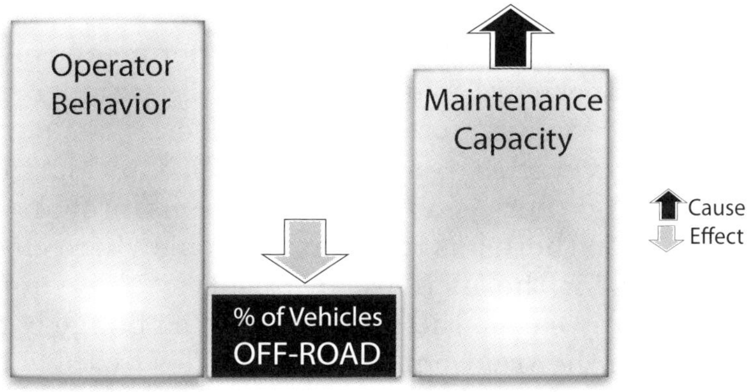


*Figure 9-3, Effect of increasing Maintenance Capacity on the percentage of vehicles off road.*

**If Operator Behavior worsens or moves negatively, as shown in Figure 9-4, there is usually a negative effect on availability, and the percentage of vehicles off-road generally increases.**

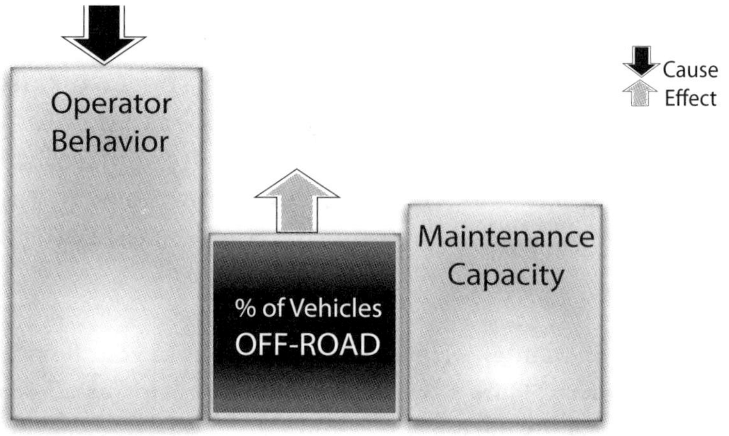


*Figure 9-4, Effect of less positive Operator Behavior on percentage of vehicles off road.*

The diagram in Figure 9-4 does not mean that operator experience is decreasing if the number of vehicles off-road is increasing. However, if the percentage of vehicles that are unserviceable is increasing, and there is no change in the maintenance system, then there has to be change somewhere else to account for the increase in unserviceable vehicles. In many cases, the cause of such a change can be found in the treatment of the vehicles by the operators — there is likely more capacity available to make improvements in the operator's behavior with regards to their use of the vehicles.

**If the percentage of vehicles off-road column is moving upwards, as shown in Figure 9-5, then both the outside columns decrease, showing that availability can be affected by a change to either or both of the other variables.**

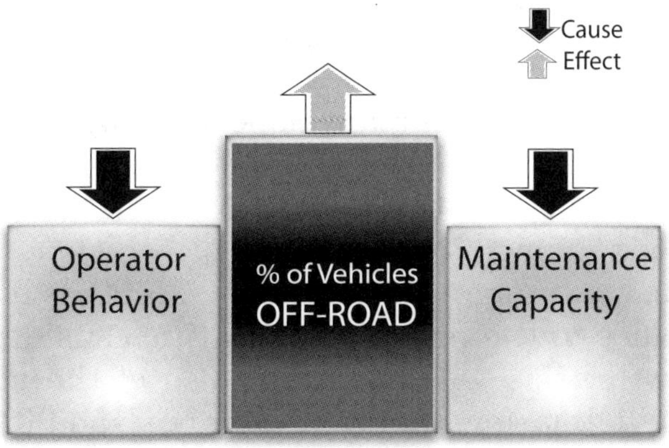


*Figure 9-5, Effect of less positive Operator Behavior and Maintenance Capacity on the percentage of vehicles off-road.*

The purpose of Figure 9-5 is to highlight that both sides of the equation are linked to the maintenance state and availability of vehicles. The quantity of vehicles off-road is not solely linked to maintenance tasks. If the percentage of vehicles off-road is too high, or increasing, the situation can in most circumstances be addressed by improvements resulting in an increase to either or both of the outer columns, which will reduce the number of vehicles off-road. Likewise, a combined approach in both the operator and maintenance areas will likely address availability issues quicker and more comprehensively than only improving one of these two variables.

# Chapter 9

**The quality of operator training, including the effect it has upon the condition of vehicles in your fleet, has been long known and must be given due consideration. The condition of the vehicles in daily use has a direct effect upon their availability. A high standard of user/operator training can thus give significant benefits to vehicle availability.**

The most important aspect of operator training is not its duration, the quantity of topics covered, or even the depth into which those topics are covered. From the perspective of vehicle availability, the key element in operator training is its direct relevance to the practical operating considerations that will arise with the daily use of the vehicles. The syllabus for operator training should, therefore, include dealing with problems that have actually occurred on the vehicles during their previous usage. Instruction on how to continue to operate a vehicle with a mechanical issue, without damaging the vehicle further, should be given to the operators. Such instruction and training should detail the on-the-spot solutions or improvised remedies and options available to drivers of malfunctioning vehicles. Also, importantly, what courses of action will worsen the state of the vehicle and which are deemed inappropriate.

Detailed training of operators on the symptoms and remedies of proven faults with a particular vehicle type will allow them to identify the symptoms of an issue when it occurs and to rectify it in a manner that will minimize further damage to the vehicle. This training will also give them a good basis to identify and temporarily rectify unknown faults, or to identify the times when it is appropriate to cease driving a vehicle or request the assistance of qualified technicians. Such training also gives the drivers a greater appreciation for the mechanical care of the vehicles and increases the likelihood that they will cease to drive a vehicle if they cannot identify or fix a problem on the side of the road.

During the invasion of the Soviet Union by Germany during World War II, the vast majority of Soviet losses of T34 tanks were due to vehicles being abandoned upon minor breakdown or running out of fuel. This led the Soviet Union to begin training their tank crews inside the production factories, where they learned vehicle engineering and maintenance before ever being taught how to operate the vehicles.

Unfortunately, many organizations have lost their appreciation of the importance of comprehensive crew maintenance training and of implementing an ethos of more than just token crew maintenance. Such training can be as beneficial to civilian fleets as to military ones. Crew maintenance should encompass limited true maintenance tasks and not simply consist of topping up oil and changing wheels. Organizations without an appreciation of detailed crew maintenance are solely dependent upon the services of qualified maintainers.

A philosophy of taking a vehicle off-road immediately upon becoming aware of an issue, and requesting a recovery or qualified technical assistance, is a very tidy philosophy in theory; it is unfortunately also very expensive. In my experience, it simply does not take account of human nature. For example, on a cold wet night when drivers of a malfunctioning vehicle are looking forward to getting home or maybe are planning to take a family holiday the next morning, they will make every effort to keep driving despite a misfiring engine, black smoke, or other problems, in the hope of getting their vehicle back to the base or depot. This decision can often cause a minor issue to become significant vehicle damage in the process. Subsequently, the answer from the drivers, upon investigation, is usually along the line of "I didn't notice a problem with the vehicle." It is often difficult to challenge this response, because the drivers have not been trained to diagnose mechanical problems.

If, however, drivers have received some fundamental training beyond what can be essentially termed *lip service*, they may have been able to tighten loose hose clips, secure a pipe, or carry out other remedial work in order to be able to continue driving and get the vehicle back, without having to aggravate the damage in the process. Drivers thus trained will also have a greater appreciation of issues that will cause significant damage to a vehicle and the times it is essential to call upon the services of a mechanic.

> *The brief training of Panzer drivers, especially in newly formed units, is reflected in the recent large increase in damaged components...at the front the drivers possibly realise a failing part of the vehicle but have no clue of how to help themselves (irregular running engine, or improper running of the radiators, etc.) This forces the driver to keep on driving the vehicle until it breaks down due to more serious damage.* *(Friedli, 2011)*

The insertion of additional, perhaps new, vehicles into an overstretched maintenance system can result in a lack of availability of these same new vehicles. A lack of availability of new vehicles under such circumstances should not always be viewed as being indicative of the reliability of the particular new vehicle type. Adding vehicles can also result in a lack of availability of any longer serving vehicles. The lack of availability of older vehicles, when additional vehicles are added to a fleet, is also not indicative of where the reliability of these older vehicles should be due to their age and expected usage versus anticipated failure rates. If a maintenance system has sufficient excess capacity when additional vehicles are purchased, the subsequent availability of both the new and older vehicles will be more reflective of their true reliability.

In an under-resourced maintenance system, as vehicles become unavailable and require maintenance, the waiting time for vehicles to undergo repairs can extend steadily. Queues for major maintenance can cause delays in even minor maintenance tasks for the vehicles that are still operational, which then causes even more vehicles to break down and, hence, further lengthen the queue for major maintenance. You can end up in a situation where vehicles are failing faster than they can be repaired. This situation is outlined in Chapter 10, Maintenance Backlog.

Determining the details of how your maintenance systems operate requires a detailed record of concrete historical and up-to-date data. The state of a maintenance system cannot be seen simply from a cursory review; it must be established from a careful evaluation of historic maintenance and vehicle availability statistics and analytics.

Your historical data must link scheduled, unscheduled, and contract maintenance costs, including labor, to individual fitted components. For example, several studies by the U.S. Army and the Congressional Budget Office have shown that vehicle age has a minimal effect upon maintenance costs. Yet it was acknowledged that this result was likely due to a lack of sufficient data on which to base the analysis. Information is key to finding those inefficiencies and faults in your system that are costing you money.

**Record keeping, accurate data, and more record keeping are of critical importance.**

Such record keeping is fundamental for the efficiency of any business operating and maintaining its own fleet. Also, having detailed analytics and data logging is not difficult with modern computer systems. It is, however, important to use a computer system that works effectively, is easy to use, and outputs the required information. A system that does not present the necessary results is of little benefit, aside from reducing quantities of paper used. Without detailed record keeping, determinations of planned maintenance and spare parts requirements are ultimately conjecture. If you do not know the frequency of component failure, you cannot truly determine, for example, how often to call vehicles in for a preventive maintenance component replacement. You are also unable to plan the quantities of components that you need to order for the following year or upcoming commitments.

In the absence of comprehensive analytics and historical data, we can still attempt to initiate a preventive maintenance strategy. This is done by carrying out replacement tasks on a conservative schedule that is loosely based upon manufacturer's recommended service and component failure intervals. Even better, we can base our replacement cycles on the same type of vehicles in operation with other organizations. The interval between component replacements can then be incrementally increased to find an ideal, depending upon how the quantity of failures during use increases. This method, however, does not address existing data collection issues nor is it as cost effective as basing your intervals on accurate data from the beginning of a preventive maintenance strategy implementation.

> *If it wasn't for some kind of PM program, every motor, pump, valve, seal, compressor, boiler, toilet, light, ad infinitum... might break down at any time.*
> *(Payant & Lewis, 2007)*

In the case of large international companies, or military organizations, accurate records of failure rates need to be maintained for each different operational climate. Failure rates in different operating zones worldwide can vary significantly depending upon the operating environment and usage profile. The workshop's management must ensure that repair data is accurately measured and recorded. This allows the information to be used for managing costs, and identifying and analyzing trends. If the management personnel in a workshop are not part of addressing any lack of accurate failure data and analytics, then management — by allowing the consequent costs to go unchecked — essentially becomes part of the problem. Without this data, and its correct usage, maintenance becomes purely reactive and is both inefficient and expensive.

Reviewing and improving the productivity of the entire maintenance system requires that data must be kept for every direct or indirect maintenance related task on each vehicle. This data should be linked to repair codes. For example, code 01 could account for all work on wheels and tires, code 02 Hydraulics, code 03 lights exterior, code 04 lights interior, and so on. This system will allow a later review of the total absorption of repair costs by each element of the vehicles in your fleet. Thus if you establish that an excessive amount of money is being spent carrying out vehicle bodywork repairs, the situation can then be investigated and appropriate action taken to eliminate the swollen costs. The individual vehicle that requires each maintenance task should be identified by its registration or hull/chassis number. In general, using the chassis number to identify a vehicle is always the preferred form of identification because it will definitely not change for the entire vehicle life.

In order to facilitate evaluation of the full productivity of a maintenance system, the following data must be maintained digitally and in an easily accessible format. If this information is maintained only in hard copy, reviewing it would involve an army of staff over a significant time period. It should be maintained *digitally,* in a format and on a system that can be *easily accessed* and used.

- ❐ **Maintenance scheduling**
- ❐ **Details and tracking of progress on particular projects such as in-service modifications**
- ❐ **Date, time, and mileage when a vehicle's status changed to Off-Road/Unserviceable**
- ❐ **Date and time when a vehicle began maintenance and identification of the individual vehicle**
- ❐ **Class code for the vehicle undergoing maintenance (type of vehicle if your fleet operates a number of different vehicle types)**
- ❐ **Mechanics responsible for this maintenance task**
- ❐ **Vehicle's odometer reading when vehicle began maintenance (recording the mileage is critical to establishing frequency of failure for individual parts at a later date)**
- ❐ **Whether the maintenance is reactive, preventive, or relating to a particular fleet project**
- ❐ **Repair type codes**

- ❒ **Date and time when each spare part was requested from stores**
- ❒ **List of spare part numbers and descriptions requested from stores for each vehicle on each job**
- ❒ **Cost of each spare part requested from stores and total costs**
- ❒ **Date and time when each spare part was picked in stores**
- ❒ **Date and time when spare parts were collected/delivered from stores to the workshop**
- ❒ **Date and time when spare parts were delivered to maintainers on the workshop floor**
- ❒ **Date and time when each vehicle completed maintenance**
- ❒ **Date and time when each vehicle was inspected or tested for completion/return to use and the result**
- ❒ **Date and time when a vehicles status changed to On-Road/Serviceable**
- ❒ **Date and time when each vehicle was collected from the maintenance facility by the operators.**

The information above will allow you to identify a timeline for your entire maintenance system. You will be able to see how long every aspect of maintenance is taking from the moment that a vehicle became unserviceable. This information will then allow you to identify choke points in your maintenance system, which can then be addressed.

It is essential that your maintenance management and data collection system be able to generate both reports and graphs based upon the historic data collected. Saving this information digitally, in addition to printing it in hard copy, will be advantageous for communicating trends and issues to higher level management.

A problem experienced by many medium-to-large organizations is that they are committed to the hardware infrastructure or to the software systems already in place, even though their information collection systems are inadequate to meet the requirements of maximizing their maintenance efficiency. This commitment is usually due to the extensive cost of switching to a new maintenance management system. In some cases, the software is integrated into other aspects of the overall business, such as financial accountancy for spare parts holdings, which can lead to additional political or departmental issues regarding changing systems.

If the current maintenance management software system does not achieve the required aims, there is little alternative to either improving it or replacing it. To do either can take investment in both time and money. But without such corrective action to implement effective data recording and analytics, reducing maintenance costs and making significant gains in maintenance efficiency will prove exceedingly difficult because it will simply not be possible to identify where the inefficiencies are located.

It is important that information on the value

or cost of any spare parts being used is gathered and tracked by your data collection systems. In some organizations, there can be a tendency for the spare parts management personnel to control all information on parts costs. Sometimes this is due simply to the organizational systems in place, but other times it can be caused by personnel acting as gate-keepers — filtering the information they communicate in order to withhold details that could be of use to them individually later. Unfortunately in most cases, the spare parts support personnel are not in a position to evaluate and alter the usage of parts during maintenance. It is, therefore, essential that the personnel directly involved with maintenance management, who make the daily decisions regarding the usage of spare parts, are fully aware of the costs involved with each individual component being used. Thus, they will be able to adapt the maintenance system to create cost savings.

As you can infer from the list above, the collection and recording of this data requires input from not only the maintenance and spares supply structures, but also the operators of the vehicles. The information collection software, therefore, needs to be integrated throughout the pre-maintenance, maintenance, and post-maintenance structures of your organization. It also has to be very easy to use, with clear screens and a logical user-friendly format, so as to minimize the time and training required for various personnel with different qualifications to enter the required data at their particular point in the maintenance chain.

The best method to achieve the required level

of data recording is for the information collection to run in the background of your ordering process software; for example, mechanics would enter the details of their required spare parts for ordering as normal from the stores. The system then stores this information in the background, without any input requirement from your personnel. The data should be recorded at the most practical level; it can be carried out across the internet if users are dispersed, or if such a system is available by using an intranet (internal network). Perhaps on a practical work level, the maintenance system for your vehicle fleet is split into different specializations, such as mechanical, electrical, communications, etc., and vehicles are queued from one stage to the next. In this case, all records must encompass the times that an individual vehicle spends not only while maintenance tasks are being carried out at each stage, but also while the vehicles are effectively sitting idle and waiting between stages.

To compile even more detailed records, some companies issue identification tags to each maintainer and spare parts supply worker. This practice allows each employee to log (or swipe) the exact times at which they started or stopped maintenance during the workday and the reasons for same. Significant efficiency benefits can be gained from such a comprehensive analysis of the maintenance system. However, this approach may be difficult to implement in an entrenched workforce or one with a strong union —it could be seen as monitoring the personnel as opposed to the tasks. Even if these tags are not possible for the reasons stated, some data is better than none; the more that is recorded and available for later evaluation, the more efficient your maintenance system can become.

Once all of the data detailed is compiled, you can identify where the maximum time lags exist in your maintenance system. Subsequent investigations, drilling down into the specific details of each maintenance stage, can then identify if delays are due to inefficiency, personnel shortages, equipment shortages, spares delivery delays, insufficient space, etc. Very interesting information can also be gained from the most basic data, such as when vehicles were declared off-road. By comparing these dates and times against an organization's training or operational commitments, you can often identify if an inordinate amount of vehicles are being declared off-road or on-road prior to these times and then seek to identify the reasons why.

If a disproportionate amount of vehicles are on-road immediately prior to intensive usage periods, it may be due to a maintenance push indicating that maintenance can be carried out more efficiently during other periods. Or it may be due to vehicles being declared on-road even though they still require minor maintenance. Such minor maintenance, if not carried out when required, can lead to major maintenance requirements and increased costs at a later date.

Most maintenance organizations have access to some level of performance and failure records similar to those described throughout this chapter. However, many large organizations have them compiled across a number of different computer and manual record types. For a large fleet, these manual records can be virtually impossible to link and analyze without a proverbial army of staff working over a significant period of time. There is no logic in the expanded effort required to maintain records if the records are inaccessible or don't record and output the required information.

A bigger consideration for any maintenance setup is the geographical distribution of facilities. If key branches of maintenance — such as mechanical, electrical, and communications — are required in a single geographical location, these branches should be located in the same compound or, preferably, the same maintenance building. In military systems, dispersal is necessary in wartime. However, from a purely maintenance efficiency perspective, modern vehicles and their ancillary equipment are integrated to such an extent that significant gains in efficiency can result from sharing the same location. These vehicle availability gains can result in fewer vehicle purchase requirements and, in turn, monetary savings. These availability gains are due to the significant reductions in maintenance time made possible when maintenance is carried out in a shared site, as opposed to work having to cease while vehicles are being moved between buildings or sites. Cross department planning also becomes more regular and effective.

When maintenance functions are located separately, personnel can become micro-institutionalized within their own workspace. But when they are located on a single site, better communication and cooperation among maintenance specializations occurs. Depending upon your local laws and organizational procedures, there can be practical difficulties resulting from unnecessary dispersion within a small or large geographical area. The availability of suitable qualified personnel to drive vehicles from one location to another is one such practical consideration. If these personnel are involved in maintenance, the time they spend transporting vehicles withdraws them from their maintenance tasks, causing delays in the overall maintenance system. The effects can be given a financial value based upon time spent on these other tasks, and the wages of the maintenance personnel added to the physical movement costs of the vehicles for fuel, etc.

Consider the following recommended items when evaluating the overall maintenance setup of your organization. All of this information should be readily available if correct record keeping has been maintained in the months and years prior to the evaluation:

- ❒ **Determine the total weekly quantities of vehicles that are serviceable or unserviceable over the necessary evaluation period, usually no less than two years. Vehicle operators/drivers should be required to keep accurate records of the exact date and time that vehicles are declared unserviceable or off-road and the reasons for same. Again, this information needs to be digitally recorded into an integrated information collection system. It is of little use if it is physically written into numerous vehicle logbooks that cannot be easily reviewed.**

## Evaluating The Maintenance System

- ❐ Identify the percentage of time that faults for each individual maintenance branch (Mechanical, Communications, Electrical, etc.) were causing vehicles to be off-road during the evaluated period. This information is even more important if branches are not located together; it can be determined by the quantity of vehicles undergoing maintenance in each branch and the quantity queued for each branch.

- ❐ Evaluate each maintenance branch (Mechanical, Communications and Electrical, etc.) to determine the causes of greater or lesser vehicle quantities being unserviceable while awaiting maintenance in each branch (backlog for each individual branch) during the evaluated time period.

- ❐ Identify ways to resolve delays, clear backlog in each branch, and carry out a cost-benefit analysis of each option.

- ❐ Ensure that vehicles are not in your workshops unless they are being worked on. The maintenance system should be turning over vehicles at a rate equal to or exceeding the required maintenance. There should not be a requirement for vehicles to sit idle at your maintenance facilities for more than 1 day before beginning maintenance.

- ❐ Cross check the dates of overall high and low vehicle availability against the occurrence of operational or training commitments so you can identify any peaks or troughs in availability.

- ❐ Identify the causes of such peaks and troughs and the benefits to be gained from any changes to maintenance or operating procedures resulting from this information.

# Chapter 9

- ❐ Identify the costs involved in transporting unserviceable vehicles and components upwards and sideways through the various levels and branches of the maintenance system. Include driver wages, fuel usage, vehicle wear, and secondary or tertiary costs such as requiring a higher quantity of transport or recovery vehicles in the fleet.

- ❐ Identify the quantity of each individual component replaced on each vehicle and the total quantity throughout the fleet during the evaluation period, including the individual and total cost to the organization. This will also allow identification of repeat failures on a specific vehicle, a specific vehicle type, a specific component, or throughout the fleet.

- ❐ Evaluate the amount of your mechanics' time actually spent hands-on carrying out repairs. Known as Hands-On Tool Time (HOTT).

- ❐ Identify ways to improve the reliability of regularly failing components to reduce maintenance costs and vehicle downtime. This may require the cooperation of the manufacturer, or your in-house research and development department if you have one.

- ❐ Check the availability of tools and equipment for carrying out maintenance tasks, to ensure there are no undue delays due to lack of maintenance equipment.

- ❐ Evaluate the cost effectiveness of outsourced work versus in-house repairs.

- ❒ **For non-critical repairs, which are unlikely to degrade into more significant damage on your vehicles, ensure that vehicles are not made unavailable for use or taken off-road until all necessary spare parts have been made available.**

- ❒ **Evaluate the quantity of repeating failures on each vehicle type in your fleet and the nature of these failures. Using this information, you can identify possible key components requiring modifications or alternatives to reduce future failures.**

## Scheduling

Scheduling is the process of planning how to put a maintenance plan into action. Planning for the plan! In any maintenance system, we must ensure that the scheduling of work tasks is given more than just a cursory thought. Without correct scheduling procedures, we will be engaged in a constant uphill struggle against inefficiency.

It is important for the overall effectiveness and efficiency of any maintenance system that there is a detailed schedule in place, and that the maximum number of maintenance tasks are being conducted according to a schedule. Accurate scheduling of maintenance is essential to ensure that vehicles are unavailable or off-road for the minimum amount of time. The amount of time they spend in the workshop must be minimized; we also need to make sure that the full benefits of preventive maintenance are being achieved. There is no logic in on-going scheduling errors or delays that result in a vehicle requiring reactive maintenance, simply because it did not come into the workshop on time to avoid a failure. Likewise, there is no point drawing up a maintenance schedule without first checking the availability of the required parts and components to complete each task. Scheduling involves reviewing the availability of all required items to fulfil a particular maintenance task, prior to a vehicle being called in for maintenance. It is also worth noting that reactive maintenance is an enemy of scheduling...you cannot plan for unplanned failures!

As with record keeping in modern workshops, the task of scheduling vehicle maintenance is made much easier through the use of computerized maintenance management systems. These systems can also provide an auditable evidence record of maintenance completion. However, the implementation of computerized maintenance scheduling is only possible if your organization is in a position to introduce such a system or integrate one into your current hardware and software configuration. This may not be financially viable, depending upon the scale of the present computer systems in place and to what level the systems are cross compatible with an addition piece of software. The current system may perhaps require wholesale replacement.

In order to ensure that scheduled maintenance tasks are conducted quickly, you must have a ready availability of maintainers, tools, maintenance manuals, and spare parts, and also use standardized procedures. Because the task is scheduled ahead of time, there is no excuse for a failure to carry out the necessary preparation, either on the part of the individual mechanics or on the part of the larger organization supporting them. The most efficient maintenance schedule will allow work to be carried out when there is a minimal demand for vehicle availability, or while a vehicle is undergoing other maintenance tasks such as routine servicing. Maintenance scheduled in such a manner to be timed as efficiently as possible can be facilitated with preventive maintenance, but it cannot be facilitated for reactive maintenance if vehicles are suffering an inordinate quantity of unplanned failures.

It is also essential that flexibility is built into the maintenance schedule, perhaps in the form of a number of unallocated hours left open each week. Such hours will allow your maintenance system to cater for unplanned maintenance tasks that arise during the week; they will also facilitate other unplanned issues such as tasks unexpectedly requiring a greater amount of time due to other maintenance requirements that are discovered after vehicles have been inspected in the workshop.

> *Surveys show that in many plants, hands-on-tools time (HOTT) is only about 25%. Delays account for much of the remaining 75%. Improved scheduling and work preparation can raise HOTT to 50-60%.*
>
> *(Narayan, 2012)*

With regards to production maintenance, the priorities for inspection and preventive maintenance are those jobs that affect the overall integrity of the plant, followed by production-related jobs. In vehicle maintenance, with regards to integrity of the vehicle, priority for preventive maintenance should be given to those items, such as brakes, that may affect the integrity of the overall vehicle platform. Brakes, although often designed nowadays to have a safe failure mode, can potentially result in the loss of a vehicle if they were to fail. In addition, if brake pads are not replaced soon enough, scoring of the brake drum can occur, thus necessitating extra repair expenses. Components that have a high operational speed or rpm, and that are not mounted within their own casing, are another example of suitable candidates to be prioritized in your preventive maintenance schedule. The catastrophic failure of these has the potential to cause significant ancillary damage to other vehicle components that are situated in close proximity.

When scheduling preventive maintenance tasks, it is important to consider safety critical tasks. Some of these, such as the brake example above, may overlap with integrity priority tasks. Safety critical tasks should be conducted within a maximum of 10% time range of the maintenance schedule. It is desirable to have such tasks conducted as close as possible to the scheduled maintenance time.

To check the accuracy of your maintenance schedule, and the efficient execution of work in accordance with your schedule, we can use the following formula. This ratio can be calculated for any time period desired, one month, one week etc.

*Execution Ratio = JC : TJ*

*JC = Jobs Completed on Schedule Within Time Period*

*TJ = Total Jobs Scheduled For Time Period*

For example, if you completed 15 jobs on schedule (JC) out of a total of 20 jobs scheduled (TJ), then the ratio would be 15:20 or 1.5:2. The closer your ratio is to 1:1, the better your maintenance system is performing with respect to your maintenance schedule. If your Execution Ratio never seems to be moving towards 1:1, then it is likely that your schedule is unrealistic with respect to your current maintenance system and resources, or improvements in your maintenance system need to be made.

Depending upon the nature of your organization's operations, the use of standby vehicles to reduce the effect of scheduled maintenance upon the overall business is also a possibility. Quite often having vehicles on standby is a waste of expensive capital resources and not financially viable for civilian companies.

The use of such a vehicle on rare occasions, although possibly having a positive effect on the company's profits over the course of a year or more, may still not offset the original costs of purchasing that extra vehicle. It may, however, be financially viable to purchase an extra heavy, longer haul vehicle, if you are doing so in place of a smaller vehicle for local requirements. The initial cost may be higher than a smaller unit and the running costs may be slightly higher, but these costs are potentially offset by the benefit of having this vehicle available to replace the larger units in your fleet, if required. Thus, a larger standby vehicle can provide a standby role for a number of different vehicle types.

The financial viability of purchasing and operating a larger vehicle than required in one role, which can fill in temporarily for other vehicles undergoing maintenance, is subject to the proportion of your bottom line being achieved by each run with your vehicles and the number of failures of these vehicles per year vis-à-vis the impact of these failures on profits. There can be significant and overriding benefits for military operations in having standby vehicles for a specific mission. In the civilian sphere, the consequences of a vehicle failure without immediate replacement availability, are unlikely to have the same level of consequences as in the military.

The most effective way to facilitate standby vehicles is to have a comprehensive preventive maintenance system in place and your excess vehicle/s rotating through the maintenance system before going off-road. The less economical practice would be having a standby vehicle sitting idle, awaiting use due to the unexpected failure of another vehicle and the consequent reactive maintenance.

It is often of benefit to have more than one schedule in a large fleet maintenance operation. The first can consist of short-term scheduling, which will incorporate a daily schedule. The daily schedule should outline the specific tasks for today and how long they are expected to take. It should allow for unforeseen circumstances that might arise that morning, such as employee illness.

An additional weekly schedule will allow for the planning of resource allocation for the following week. This is compiled at least a week prior to its start date, while the workshops are operating from the present daily schedules.

Another required schedule is a longer-term schedule compiled to plan tasks for, at a minimum, the upcoming month. It is desirable to plan up to 6 months in advance, if possible. The length of time which your schedule covers depends upon the regularity of your fleet's operations, but it can include all regular service tasks and expected preventive maintenance. It should be checked and extended on a weekly basis, which will give a minimum of three weeks prior notice for upcoming work. Finally, your longer-term schedules should always leave a larger percentage of unallocated time than the shorter-term schedules — a larger quantity of unforeseen circumstances can occur over the greater time period being scheduled.

## Life Cycle Costs

The term *life cycle costs* is an umbrella term that is fundamental to modern fleet operations, yet is ultimately made up of the various individual costs that arise throughout the entire life cycle of your vehicles. These individual costs are unique to each vehicle and fleet profile. Cost Engineer Dean Fanning describes the essence of life cycle costing quite clearly. He outlines that it is comprised of point estimates of costs, a discount rate, and a sensitivity analysis, observing that,

# Chapter 9

*...just because something is less expensive to start with (capital cost) it does not mean it is the best economic alternative.*

*(Fanning, 2014)*

When purchasing a vehicle, we must consider not only the capital expenditure on a new unit, perhaps $150,000, but also the costs that the vehicle will generate until the day it goes for disposal or stops being used in our fleet. Vehicle life cycle costs are those expenditures incurred throughout the vehicle life that include items such as the total cost of fuel and the total maintenance expenditure. Although it is important to estimate these when selecting which vehicle to purchase, it is also important to remember that these costs are not static. As a vehicle ages and becomes less efficient, the ongoing running costs will increase. Therefore, the profile of your estimated life cycle costs will be higher due to increased operational costs in the later years of a vehicle's lifespan than in the early years. Yet, the increase in these costs is not always as significant as we might assume because ongoing vehicle maintenance involves replacing old, inefficient, or well-used components with new components. Thus, maintenance can somewhat counterbalance the otherwise intuitive linear increase in costs we would otherwise expect.

When purchasing and operating vehicles, there can be very significant differences between vehicles, even vehicles of the same overall type, in relation to total life cycle costs. For example, one vehicle from manufacturer A may have significantly faster maintenance turnaround times, but this saving may be dwarfed throughout the vehicle's anticipated lifetime by another vehicle design from manufacturer B with much greater fuel efficiency, another with less expensive replacement parts, or another with much higher reliability during its usable life.

We need to consider both costs that are individual to a vehicle or accumulated across a fleet, depending upon the nature of each cost type and the overall fleet profile. In a mixed fleet with a proportion of older vehicles that are fuel and maintenance intensive, we may also have a proportion of newer, more efficient vehicles. The impact on our ongoing expenditure and balance statement can be improved or worsened significantly by the life cycle costs of our different vehicle types or their age. Thus, the average life cycle cost of our entire fleet is affected positively or negatively by the fleet profile. Our ongoing expenditure may be driven upwards by inefficient vehicles to a greater extent than our selection of new fuel efficient models are reducing our costs.

We must never accept the life cycle costs estimated at vehicle purchase as being set in stone for the total vehicle life. New technology, fuel additives, component modifications, and other options can be implemented in a mid-life upgrade program or life extension program. These upgrades have the potential to reduce our ongoing expenditure later in the vehicle life cycle. As a consequence, we might reduce the total costs incurred throughout the overall life cycle of our entire fleet.

Reducing your total life cycle costs is quite simply a matter of reducing the individual costs that arise intermittently or repeatedly during the vehicle lifetime. Reducing these costs is achieved by processes such as reliability engineering, improving maintenance efficiency, reducing running costs, and reducing ancillary costs, as discussed throughout this book.

It is very important to consider life cycle costs when initially deciding upon which particular vehicle type to purchase for your fleet. By carefully calculating the total life cycle costs for each potential new vehicle type, we can ensure that we identify the vehicle with the highest likelihood of providing the greatest financial return over its service lifetime. And this may not be the same as the vehicle with the cheapest initial capital purchase cost.

Dean Fanning states that life cycle is about relative values between alternatives; if there is only one alternative, then no life cycle cost analysis is required. In the case of vehicle fleets, I would argue that the second part of this assertion is not true. For example, there may be only a single vehicle type that you can purchase to meet a required need. However, if you fail to conduct any life cycle cost analysis on this single vehicle option, you cannot estimate the viability of the vehicle to enter our existing fleet or the effect it will have on the company's cash flow, balance statement, and maintenance budget in the medium-to-long term. Having conducted a life cycle cost analysis on such a vehicle, we can then consider other factors, such as our overall fleet replacement program, to ensure that any new vehicle does not tie up too much capital at the same time that we plan to require cash on hand for other purposes in the future.

CHAPTER 10

# MAINTENANCE BACKLOG

Backlog in maintenance is a key indicator that you can use to measure the work flow and productivity of your workshops. Hence, it is given its own chapter here. Backlog cannot be ignored or it will often steadily increase, creeping up on you until it is at a point where it demands attention.

We previously mentioned the possibility of maintenance queues becoming ever longer as the queues cause the condition of the vehicles that are still functioning to gradually degrade. Essentially this means that, when queued, maintenance is allowed to build up beyond regular levels into a backlog. The backlog itself can then compound any factors that initially contributed to its existence. In turn, an ever increasing quantity of vehicles go off-road accompanied by ever increasing difficulties in maintaining the necessary numbers of available operational vehicles, until the situation becomes critical. Thus the maintenance queues become self-antagonistic. This cumulative effect of increasing failures can occur over a period of years. In some cases, a number of significant negative contributory factors occur simultaneously, and the backlog is generated within a much shorter period of time, such as weeks or even days.

As maintenance queues become longer, the use of the remaining in-service vehicles also increases, to maintain outputs, causing more wear and higher failure rates on these remaining vehicles. These higher failure rates then add further to the burden of the overstretched maintenance system, further increasing maintenance delays for queued vehicles, as well as the quantity of queued vehicles. Thus, a cycle of perpetual delays begins in the maintenance system. This situation can also lead to a skewed reflection of vehicle reliability for both new and older vehicles, and a skewed perception of the operation of the maintenance facilities.

The turnover of vehicles in a maintenance facility is always linked directly to a number of factors — the number of maintainers available, the physical size of the facility, and the spares LTTA (Lead Time To Availability) from the external suppliers and locally within the organization — and to the availability of the necessary equipment allowing maintainers to carry out their work. These factors directly define the number of vehicles that can be maintained at any given point in time. They also define the number that must be placed in a queue awaiting maintenance. If there are too few maintainers, or nowhere for them to work, vehicles will not be turned over efficiently. If the spares supply system is not rapid and efficient, vehicles will not be turned over efficiently.

From a larger organizational perspective, the requirements necessary to relieve the increased burden of maintenance backlog, combined with any new vehicle requirements and on-going maintenance, can often be greater than the initial maintenance investment requirements that would have prevented such a situation from arising.

Unfortunately, there is no magic wand to address a maintenance backlog. It must be dealt with by meeting the appropriate maintenance requirements — such as increases in space, equipment, and staff — in order to increase the turnover rate of vehicles. Larger spares supply requirements must also be met to facilitate an increased maintenance turnover, which may also require increases to the levels of support staff. Additional work hours may have to be introduced in the form of maintenance intensive periods or overtime facilities.

Avoiding or minimizing these staff and equipment increases is sometimes possible with a detailed analysis of the workflow in the maintenance facilities and the spares supply system. Using such an analysis, it may be possible to identify ways to increase the vehicle turnover inside current resources. However, without such an analysis to identify inefficiencies and quantify possible gains, any investment can easily be misdirected and is not advised.

Inefficiency in spares supply can be offset, to a limited extent, by excess space capacity in the workshops. Excess capacity allows maintainers to

**In 1941, at the beginning of the WWII operation Barbarossa, German vehicles were not being maintained properly, nor were they being supplied properly. In one case, during August 1941, 103 Panzers were lost due to spare parts shortages and a lack of scheduled maintenance. Only 59 vehicles were lost due to enemy action during the same period, which gives us a clear perspective on the effect that maintenance system failures can have on our fleet.**

begin a second job while awaiting spares for the first; it also supports efforts to address any maintenance backlog. Such a practice however, depends upon the second vehicle requiring different spare parts than the first, and these parts must be immediately available. It is also an inefficient use of labor and workshop space, which causes its own increase in individual vehicle maintenance time because maintainers have to reacquaint themselves with various jobs as spares become available.

Each mechanic must first be a diagnostician, responsible for diagnosing and identifying not only each problem on a vehicle, but also its immediate and ultimately root cause. If there are insufficient quantities of specialist diagnostic and repair tools and equipment for the number of staff, vehicles will not be turned over efficiently. This is often the case with specialist tools, which can be expensive. By virtue of their title as *specialist*, there can be in some instances an inclination to believe that not all mechanics require a full set of tools. In many cases, that may be true, but it is dependent upon the size of the fleet and the ongoing maintenance demands.

In order to avoid an availability-driven maintenance organization, which limits your turnover of vehicles, you must ensure the ready availability of all necessary tools and equipment for your total quantity of staff. Having the best tools available for your workforce is a key role of management; it's a sign of a professional and effective organization.

Critical backlogs can result from a number of different circumstances and factors. A number of potential causes are listed below. More than one of these circumstances arising at the same time, will compound the situation.

- ❒ **Insufficient number of employees**
- ❒ **Employee illness**
- ❒ **Employee leave (Approved or Absence)**
- ❒ **Industrial action (Official or Unofficial)**
- ❒ **Maintenance equipment breakdown**
- ❒ **Poor maintenance planning**
- ❒ **Lack of care by vehicle operators**
- ❒ **Increasing vehicle usage**
- ❒ **Increasing vehicle age**
- ❒ **Delays in spare parts availability**
- ❒ **Insufficient workshop space**
- ❒ **Increase in fleet size**

Unexpected illness can result in an increasing number of vehicles being queued for repairs. This situation can be especially pronounced if an unavailable employee is a specialist maintainer, of which there are limited numbers presently employed in the organization. Even worse — if an unavailable employee is the sole specialist employed for a particular type of maintenance task. Automotive electricians can often fit into such a category because they will be employed in lower numbers than general mechanics. In military organizations, there are often a variety of other specializations necessary to maintain a complete platform. In a situation where the required maintenance staff are unavailable, the sub-contracting of work out to an external company is often the best or only short-term option.

Most maintenance systems in many industries internationally are currently operating on ever tighter margins and decreasing personnel numbers. Any failure to manage an increase in vehicles awaiting maintenance, either through external contracting of work, or replacing a missing employee on a temporary basis, can result in a minor temporary backlog becoming a severe backlog. This is not only expensive to rectify, but the longer such a situation persists, it will have a cumulative effect on the bottom line and operational effectiveness. Most organizations in the current economic climate are unlikely to have sufficient personnel numbers to absorb the extra workload from unexpected illness of a key maintenance specialist.

The taking of leave by maintainers can have similar negative consequences for vehicle turnover. Shortage through leave has an advantage over shortage through illness, in that leave can and should be pre-planned and scheduled in such a manner as to minimize the overall effect upon queued maintenance. If the organization has fully implemented a preventative maintenance philosophy, then the schedule can be correlated against the leave schedule.

The breakdown rate of maintenance support equipment should be reduced through the full implementation of a preventive maintenance schedule for all such tools and equipment. This area is often overlooked by many fleet operators, but plans to reduce the effect of major support equipment failure should be put in place prior to such a situation arising. Alternative arrangements, such as having backup units for less expensive tools or planned rental arrangements for the more expensive units, should be in place ahead of time.

Among the first indicators of an increased lack of care by vehicle operators, or of increasing vehicle usage resulting in increased maintenance burdens, is an increase in unplanned or reactive maintenance, or shorter intervals for planned maintenance based upon variable parameters such as mileage. These changes will cause the quantity of backlog and queued vehicles to rise. Delays in the supply of spare parts due to fundamental availability issues caused by unplanned maintenance, or scheduling errors for planned maintenance, can also cause an increase in the quantity of queued vehicles. It will be very difficult to evaluate the scale of any backlog increases without accurate historical data on failure and maintenance turnover rates. The absolutely essential nature of maintaining such records is a recurring theme throughout this book.

One of the most dramatic generators of maintenance backlog is industrial action. An industrial action can range from a minor dispute at a local level, causing one or more employees to slow their work rate, to a full scale strike which, in a short period of time, can completely shut down a fleet. The consequences of such actions must be thought through ahead of time. The response to take on these events, both predictable and unpredictable, must be in place in the form of a written plan developed before the events occur in order to avoid the worst effects upon your organization. Logistics and the movement of goods can be thought of as the arteries of modern business organizations. Any interruptions will have dramatically negative effects upon the whole. Having a proactive plan and an efficient maintenance system in place to avoid backlogs is, therefore, essential. These plans should be constantly reviewed for improvements and to identify minor problems before they escalate.

In *Army Logistician*, Capt Vincent R. Lindenmeyer and 1st Lt Gilbert J. Duran describe an effective system of reducing backlog that was implemented in the 782nd Main Support Battalion of the 82nd Airborne Division. They implemented a program called the Heavy Maintenance Intensive Week (HMIW), which was conducted on a quarterly basis. This involved 100% of the maintenance capacity working two 11-hour shifts over a period of 1 week. There was a 4 1/2 hour shift overlap. The first shift would focus on completing unfinished tasks, preparing new tasks, and notifying customer units that vehicles or equipment was ready for collection. The second shift, which ran primarily outside normal hours, could work *"unhampered by the normal duty-day interruptions."* This system gave tangible results very quickly. During the first week-long HMIW, the personnel completed 100% of the jobs that had been on hand for more than 30 days, and 44% of the jobs that had been on hand for less than 30 days. After the second HMIW, all workable backlog jobs were completed and the total backlog was reduced by 84%.

> *Developing a proactive plan of action to support its customers during periods of high operating tempo and to reduce maintenance backlog during critical times of the division training cycle meant the difference between excellence and mediocrity.*
>
> *(Lindenmeyer and Duran, 2001)*

If your organization has built up a maintenance backlog, the implementation of a temporary system to increase turnover is the most viable way to eliminate the backlog and return to normal operations. Doing so may involve the payment of overtime rates or time in lieu to the necessary employees. If the backlog is seen to be steadily increasing in your organization, it is likely that the situation has reached the point where the backlog has become self-generating due to extensive maintenance delays and will become critical without decisive action.

The ideal situation is one whereby preventive measures are in place on an on-going basis, so that backlog never reaches the point where excess working hours are required to resolve it. Unfortunately, there are often unexpected circumstances that can cause maintenance backlogs and, in accordance with Murphy's Law,

*"Anything that can go wrong will go wrong."*

such maintenance-generating situations will tend to occur at the worst possible times in the operational or maintenance cycle. As such, preparation and forward planning for the unexpected are key to minimizing the impact of these situations as they arise.

CHAPTER 11

# PURCHASING ADDITIONAL VEHICLES

Although it is possible to keep a vehicle operating indefinitely by the replacement of components ad infinitum, it is expensive to run older vehicles in a fleet. Due to mechanical inefficiencies, direct running costs increase with age. Indirect running costs, such as a requirement for more regular inspection and increasing numbers of maintenance personnel to cater to more frequent breakdowns, also increase. There comes a time in any vehicle life when you must consider the viability of repair versus replacement.

Hidden costs are also incurred due to inefficiencies with older vehicles in your fleet. Newer, more efficient, and reliable models are more cost effective, and can reduce an organization's overall outlays. A number of straightforward considerations should be taken into account when deciding whether to repair a damaged or unserviceable vehicle or to simply replace it with a new model.

The decision to replace a vehicle in service should consider not the initial purchase value of the present vehicle, but instead the present residual value, accounting for depreciation. This is the actual value at the time of the necessary repairs; it accounts for the vehicle's age and anticipated remaining life. If the repair costs for a vehicle exceed 60–80% of the *remaining depreciation* on that vehicle within its anticipated life, it is not economical to repair it. This method (versus using 60–80% of the total remaining depreciated value) has some built-in flexibility to allow for the expected reduced scrap or resale value of a vehicle that has had major repairs, rather than calculating your percentage based upon the total remaining depreciated value at the time of repairs.

Consider, for example, a box body truck with an anticipated lifespan of 15 years. Because it appears tidy and allows for neat accounting, a fixed period age is widely used by many organizations to calculate vehicle life. Normally, vehicle lifespan should be measured using mileage. Although lifespan measurement using years allows us to compare productivity reductions relative to newer vehicles, using a fixed time period such as years is not recommended in practice. It is too simplistic and does not account for annual mileage rates or different engine and component life spans across different vehicle types.

Let's continue with our example, using a 15-year lifespan. If your vehicle is currently 12 years of age, it has 3 years of operational life remaining. If the current value of the serviceable vehicle is $38,000, accounting for depreciation, and the depreciation over these 3 remaining years is expected to total $12,000, then the repairs are uneconomical if they exceed $7,200 (60% of the remaining $12,000 depreciation). This amount may seem low, but the viability of the repair costs must be based upon the ability to recoup them within the vehicle's remaining lifespan. By using this method, you can ensure that the cost of repairs can be recouped within the remaining depreciable life of the vehicle.

If we had based our viable repair cost on the total remaining depreciated value, it would have been 60% of $26,000 ($38,000 – $12,000). This calculation would give us a viable repair price of up to $15,600, which would be an inordinately large cost to recoup within 3 years on a vehicle with a resale value of only $26,000.

In addition to this basic accounting for remaining depreciable value, other considerations must be taken into account when considering repair versus replacement: the vehicle type and ease of maintenance. The costs of keeping a vehicle type running, if it is no longer widely used in your organization, may make repairs uneconomical. This is due to the ancillary expenses of maintaining expertise, equipment, and spares for a platform that provides a limited role in your organization, or perhaps is being withdrawn from service. There may be efficiency gains to be made by replacing the unit with a newer vehicle; these gains can often compensate for any minor losses incurred through retiring a vehicle early. Some organizations depreciate their vehicles' full book value over the first few years of operation partly for this purpose. If a decision is made later to replace a vehicle early, its book value is already zero.

The purchase of a new vehicle fleet, or additional vehicles to complement an existing fleet, always brings a number of maintenance-related considerations that can have a significant effect upon the performance of not just the new vehicles, but all vehicles in service. Considerations surrounding the capacity of the maintenance system and the effect of new vehicles entering the system, such as those discussed in Chapter 9, cannot be simply ignored or put off until after the vehicles' entry into service.

Prior to embarking on any purchase program, you must establish the capacity of the present maintenance system. Fully evaluate the areas described in Chapter 9 and estimate the maximum capacity should each of the identified delays and failings in the system be resolved. Identify the static factors that cannot be changed. These static factors may include the number of employees or the physical capacity of the workshop buildings. The increase in vehicle turnover with respect to changes in each identified area should be estimated. Once complete, an evaluation should be conducted to identify the changes that are viable to make to the maintenance system and the total estimated benefit these changes will have on vehicle turnover vis-à-vis the cost of implementation. Again, it is not viable to conduct this evaluation if your organization has not maintained accurate and comprehensive maintenance records in an accessible format.

If your organization has a constant queue of vehicles awaiting maintenance, there is likely potential for improvement to the availability percentage by reducing or removing this queue. Implementing improvements to maintenance turnover can generate such a reduction, and the size of this potential reduction can also be established. If this potential improvement results in a sufficient quantity of vehicles, now fully serviceable, to meet your company's needs, it may negate or reduce the requirement to purchase additional vehicles. If, however, the quantity of vehicles available is still too few to meet requirements, or there are no significant ways to improve the maintenance turnover of vehicles, then additional vehicle purchases should continue to be investigated.

It is important to note that the expense of additional purchases is incurred *after* reviewing the possibility of increasing vehicle availability in other ways. Additional vehicle purchases should not be the first option when availability issues arise. Inserting additional vehicles into a system that has not been evaluated may compound maintenance-driven availability problems as opposed to resolving them.

The number of vehicles being turned over in a maintenance facility is directly related to the number of maintainers. Without an increase in the number of maintenance personnel, the improvements possible due to other changes are often limited. However, an increase in maintainers should only be undertaken after all other failings in the system have been addressed. First pursue other methods of addressing maintenance capacity issues.

As a rule of thumb, the maintenance requirements of an organization will increase as a straight line graph proportionately to the quantity of vehicles in service. This may seem obvious, but in my experience many organizations allocate capital funding for new vehicles to address availability issues, but fail to consider the consequent increased burden on the maintenance infrastructure. This relationship is illustrated in Figure 11-1.

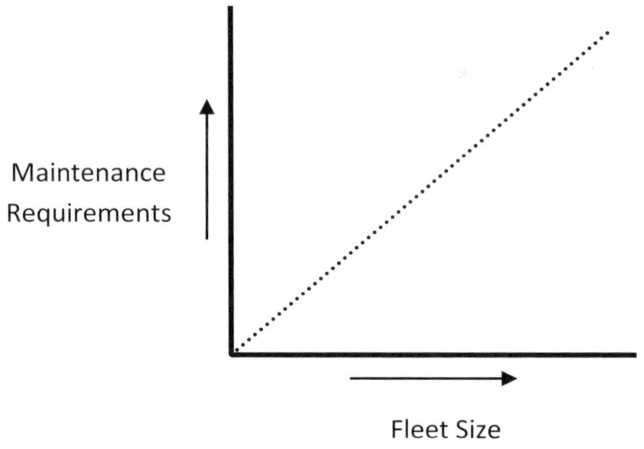


***Figure 11-1, Increase in maintenance requirements proportionate to the number of vehicles of same type and usage in service.***

This rule of thumb applies if the additional vehicles are the same type as those already serving and having a similar usage profile. Different vehicle types incur different maintenance burdens.

Suppose you have a fleet of 200 vehicles and a maintenance staff of 20. Increasing your fleet to 400 of the same vehicles with the same usage profile will result in twice the number of vehicles off-road at any given time and will require twice the number of maintainers (40) in order to maintain your vehicle turnover and On-Road state at its current level. The larger the fleet size becomes, the more accurate this average state. The required increase in maintenance staff, however, can be reduced if the inefficiencies present in the system can be identified and resolved, as previously discussed.

# Chapter 12

# Types of Maintenance Systems

The maintenance system employed in your organization is essentially a process that is followed step by step for every vehicle. It starts from the point the maintenance process is initiated while the vehicle is still located with the operators or end-users. The vehicle then goes through the maintenance system, which ends when the vehicle has reached the final point, where it is being collected by or returned to the operators.

The cost effectiveness of your fleet and the total percentage of vehicle downtime can be significantly affected by your organization's choice of maintenance system and the effectiveness with which it is employed. An ineffective system, or one that does not cater for unexpected increases in demand patterns, can lead to a failure to meet the needs of the business. In the military, it can even be a direct cause of mission failure.

In a functional approach, we can judge the success of our maintenance system by whether it has fulfilled its function during different demand periods and levels of requirement. For example, if we cannot turnover enough vehicles during a period of high maintenance demand, or the quality of the maintenance output is contributing to increased failure rates, then we can state our maintenance system is not fulfilling its function and requires review.

In World War II, a maintenance system was developed known as 3-*Level Maintenance*. This system depends upon all tasks being carried out at as low a level as possible within the maintenance structure. When a task cannot be carried out, the unserviceable vehicle is passed back along the logistics chain to be repaired at the next higher level. Either the skills, equipment, or both may not exist at the lower levels of the repair chain. As vehicles are passed through the system, the investigation and evaluation processes are repeated. The repetitive nature of these processes is an inefficient use of time and resources. Increased costs and logistical requirements are incurred when unserviceable vehicles are transported between links in the repair chain. Significantly increased downtime is also a result of this system. A number of large civilian and particularly military organizations worldwide continue to employ a 3-Level maintenance system.

> *The time spent sending equipment from one activity to the next created longer repair cycles, lower operational readiness rates and, consequently, decreases in combat power. (Astphan, 2005)*

The 3-Level maintenance system does have a number of its own benefits. One significant benefit is centralized control of fleet maintenance as you move up through the maintenance levels. When operating correctly, this feature should ensure that centralized management locations will clearly see the extent of significant vehicle faults because the vehicles are all being sent back to these central locations for repair. Tracking the rates of major failures and warranty issues becomes easier because the 3rd Line management are directly addressing these issues at their own location. Liaison with external suppliers and manufacturers is also easier because the management is directly aware of the nature of these faults.

In smaller maintenance structures with a large geographical distribution (for example a few hundred vehicles disbursed nationwide), this system can centralize the familiarity with less common component failures in the hands of an individual manager or a limited number of maintainers. In a military organization, this is a double-edged sword that requires a short explanation of centralization and decentralization.

The term *decentralization* has been much abused worldwide in the past decade, in so far as it applies to the physical relocation of personnel and an organization's facilities. This fundamental misunderstanding of the decentralization concept can lead to inefficiencies and increased costs by physically spreading out the departments of large organizations.

The term decentralization is correctly used when it refers to the dispersal of knowledge throughout a large organization to avoid the proverbial *house of cards* that may collapse if key personnel and their knowledge are lost. However, rather than centralized maintenance management, control can also be exercised through a system of repair time evaluations, inspections, and information dissemination to the lower-level repair facilities.

An alternative to the WWII era 3-Level maintenance system, previously proposed by Brig. Gen. Thomas Dickenson, Chief of Ordnance, U.S. Army, is the concept of Fix-Forward/Replace-Rear. In this system, all the necessary skills and maintenance resources to keep vehicles operating are provided at the operational maintenance level in every location. This allows the vehicles to have only one port of call in the repair chain: the local workshop. A vehicle no longer has to be passed to another level to return it to service. The local maintenance facility can carry out all maintenance up to 3rd Line/Level (1st – Crew Maintenance, 2nd – Service Tasks, 3rd – Removal and Replacement of major components). In the Fix-Forward/Repair-Rear philosophy, the 2nd and 3rd Lines are effectively merged into one functional level. Detailed repair of components is also not conducted at the local level where maintenance is confined primarily to replacement with a view to returning vehicles to service as quickly as possible. Failed components are then shipped back along the chain for evaluation and centralized repair.

In the old 3-*Level maintenance system*, local repair facilities are often required to collect new spare parts at designated intervals from centralized spares holdings. This system remains effective in a fix forward/replace rear system. However, the savings in ancillary repair costs are made by no longer requiring the movement of disabled vehicles and similar expenses.

Other considerations in the fix-forward/replace-rear system are the quantity of vehicles being repaired at each local location. If the local operators are required to maintain a very small quantity of vehicles, then the skills of the maintainers may not be kept up to the highest standards simply due to an insufficient number of different major failures appearing on each vehicle type. Also, there is an initial expense involved in upgrading local maintenance facilities and providing equipment within each one for the full maintenance of each vehicle type. These expenses can however, be recouped over future years due to increased on-road time and reduced logistics or ancillary costs.

The logistics expense of transporting individual failed components to a central location, where they can be repaired or sent for repair, is negated by sharing the repairables collection facility with the spares distribution hub. The vehicles that are travelling to collect new spare parts can be used to return unserviceable components at the same time.

> *Current developments in the design of military equipment would support the following readjustment of maintenance priorities. Discard selected components at failure (First Line); repair forward by replacement (Second Line); evacuate for repair (Third Line).*
>
> *(Cunningham, 1995)*

Some advantages of a 2-Level system, as proposed by General Mitchell H. Stevenson *(Army Logistician 2002)*, are:

- **Reduced logistics footprint in the battlespace**
- **Faster returns of equipment to the fight**
- **Decreased equipment evacuation requirements**
- **Increased productivity of maintainers, and therefore increased combat power**
- **Possible force structure savings**

General Stevenson also outlined some disadvantages of the echeloned system of maintenance as currently seen in operation with many organizations:

- **Contributes to a large logistics footprint**
- **Relies on evacuation systems**
- **Has a built-in overhead burden at each echelon**
- **Requires maintenance support for the units that provide maintenance at each echelon**

During the first decade of the 21st century, the United States Army Ordnance Corps began transforming to a 2-Level maintenance system. This transformation was carried out to make combat units more mobile and self sufficient. It was also seen as an answer to increasing strategic, combat, and technological challenges.

Making a decision to discard selected failed components at 1st line repair locations should involve a careful analysis of failure rates; it must reflect the cost effectiveness of transporting particular failed components for repair. A large number of nations and states are subject to ever more comprehensive requirements related to the disposal or recycling of various materials used in the construction of vehicles. The costs involved in disposal at more widely distributed and numerous first line locations must be considered.

A significant matter for concern regarding the 2-Level system is a risk that local pressures on maintenance personnel will result in increased cannibalization of vehicles in order to meet short-term local commitments. This practice must be countered using a detailed auditing and inspections regime from a higher level, with decisive action taken to prevent reoccurrence of such a situation should it arise. The capital value of modern vehicles is simply too high to justify cannibalization of anything other than very old units. Such cannibalization is indicative of problems within a maintenance organization; the system should be evaluated to determine why such actions are occurring. Cannibalization of vehicles due to long spares LTTA, or similar challenges, can result in a situation of increasing maintenance backlog. In this scenario, the unavailability of vehicles progressively worsens the condition of the remaining units, effectively becoming a negative loop that causes ever more vehicles to go off-road.

In my opinion, the old WWII echeloned maintenance system is inefficient. It creates longer repair cycles due to the time taken for equipment to move between levels in the maintenance chain and repetition of diagnostic maintenance activities. It also lowers operational readiness in comparison to a more streamlined 2-Level system.

In *Army Logistician*, Richard Price also discusses the concept of modular component replacement by vehicle operators. Although this would be extremely efficient and operationally desirable, it is not workable from a cost perspective without a fix-forward/repair-rear maintenance philosophy being firmly in place beforehand. The costs of replace and discard systems for a large fleet can also prove to be prohibitive and wasteful. To avoid inappropriate replacement intervals and the associated wastage, a very clear delineation must be established of which components will be sent for repair to be reused again later, and which will be discarded permanently for each vehicle type in service.

## In-House Manufacturing

When an organization has its own in-house maintenance, the effective functioning of a large modern fleet of specialized vehicles (either in the civilian or military spheres) can often lead to manufacturing tasks being carried out by the maintenance facilities. These tasks—metal working, welding, painting, plating, etc.—often result from a requirement to add minor modifications to vehicles such as storage boxes, steps, or similar items.

However, many organizations do not fully recognize this role or what is required for it to be successful within a system primarily based around vehicle maintenance tasks. Maintenance tasks will generally vary from job to job and can maintain a level of interest for personnel. Manufacturing tasks, by their nature, are repetitive and require a mindset of process optimization to produce units efficiently in bulk. The effect on personnel of a switch from a maintenance role to one of manufacturing must not be ignored, as keeping personnel motivated to work through monotonous repetitive tasks can mean the difference between success and failure. The skill set available within the organization must also be appropriate for the design of vehicle parts, both to carry out the necessary engineering tasks and also to ensure any external vehicle changes do not fall foul of relevant legislative requirements for the subsequent operation of the vehicles.

Energy usage and environmental legislation for the manufacturing process of even small quantities of items can drive costs upwards. Design and manufacture requires specific in-house skills; there are likely to be higher efficiencies present in a dedicated manufacturing organization outside your company. Also, outsourcing manufacturing related tasks enables us to effectively off-load the legal and Health and Safety responsibility for design errors. When these considerations are taken into account, sub-contracting manufacturing tasks may prove to be a better route for most fleet owners. If the entire fleet is equipped with a new additional item at the same time, then the cost of sub-contracting the design and manufacture can be a one-time expense. This approach may be more economical than subcontracting on-going tasks such as true maintenance. Of course, other considerations such as labor usage, processes, and capacity will also have an effect on costs if your organization requires in-house manufacturing. These items will have a greater effect on cost effectiveness than energy or environmental concerns on a daily basis, but all must be optimized to reduce costs.

The 12 steps involved in optimizing the inhouse manufacturing process are listed below. A more extensive discussion of process optimization is available in the *Process Optimization Guide for Military Manufacturing and Maintenance Facilities (Lin, 1999).*

Many organizations are not fully appreciative of the numerous small manufacturing jobs that are conducted on an annual basis to benefit their fleet — from production of difficult-to-source vehicle components to the manufacture of items that would otherwise have a long LTTA. It is not in the best interests of any business if funding for these essential tasks must be found through creative accounting in other budgets. It is essential that manufacturing tasks in a maintenance facility are recognized, quantified, and budgeted for accordingly.

## Types of Maintenance Systems

- ❐ Identify critical cost issues...that adversely impact operation efficiency, cost, energy, environmental, and overall performance.
- ❐ Conceptually model the existing manufacturing and maintenance operations.
- ❐ Financially analyse the manufacturing and maintenance process to develop the total cost of....critical issues.
- ❐ Develop a Process Flow Diagram to quantitatively define key technical values and costs.
- ❐ Develop a weakness analysis to challenge the existing process.
- ❐ Identify solutions to critical cost issues.
- ❐ Select the best solutions that offer the greatest savings potential and best chance of implementation.
- ❐ Develop ballpark economics for the best solutions to estimate payback.
- ❐ Categorise and group solutions as to ease of implementation.
- ❐ Conduct a formal debriefing of results to obtain management buy-in.
- ❐ Document all results.
- ❐ Propose an Implementation Plan and secure commitment to proceed.

# CHAPTER 13

## PERSONNEL

The training of personnel is important for improving staff skills and for continuous development. It is also essential for working with systems that have safety considerations, such as refrigerants used in vehicle air conditioning. However, when purchasing new vehicle types, it is important we don't place too much faith in the process of supplier or vendor training. Suppliers will often train your operating personnel in how to use equipment when everything is working. Unfortunately, they will not train them in how to resolve unexpected situations when equipment is not working properly. Unlike the purchasers, the manufacturers of military or specialist vehicles rarely use their own vehicles in large numbers on a daily basis over long periods of time. Likewise, training in-house maintainers to become familiar with the majority of vehicle faults can take years. This degree of training will not be covered on a supplier course during a period of a few weeks. In fairness, many of the vehicle faults we have to address with new vehicles are unforeseen failures that occur after 5000 km.

Staff are a key asset in any business or organization. When it comes to collecting information on vehicles being maintained, the feedback of the mechanics who are actually performing the hands-on maintenance is often the most important source available to the maintenance manager. There must be a system structured to collect the mechanics' comments and review them in an appropriate manner.

In many organizations, this process is facilitated through weekly or monthly sessions commonly called *tool* talks. These meetings must have a tight agenda defined beforehand; otherwise, they can end up taking hours to complete, or proceed toward unplanned topics.

Although staff ultimately is usually *the* key asset in a company, especially of any maintenance endeavour, they are also often a complicated and difficult aspect of business to deal with from a management perspective. Human nature being what it is, personnel will have good days and bad days, energetic and slow days, motivated and de-motivated days. Entire books have been written covering personnel management. Seemingly insignificant items, as simple as job titles, do matter. Staff need to be treated with dignity and respect, while being effectively organized by management, who are in charge and who act fairly, without any positive or negative bias, and without trying to be the staff members' best friend. This approach can be summarized by three Fs: Fair, Firm, and Friendly.

We previously mentioned that the possibility may exist to assign an identification tag to individual staff members to facilitate tracking the time taken for different jobs and tasks. Ironically, the implementation of a system to track labor time can, in some cases, actually improve morale. It will allow the mechanics to have a definitive set of values against which they can measure their work. If they meet these values, they know they are doing a good job. Unfortunately, one of the traditional aspects of vehicle maintenance is that the staff do not get accurate feedback on their work. A main reason is that there is no defined time standard in place for carrying out particular tasks. If such a system can be implemented, it will potentially have benefits for the staff and organization as a whole.

There is a definitive value allotted to staff in any organization. However, it can be difficult to measure and allot the value added by staff wages to the turnover of maintenance tasks. This value is especially difficult to measure for floor supervisors and management whose role cannot be allocated to a single individual job. However, everyone ultimately has their own job to complete, and most people realize in the current economic climate that doing their daily work as required is the minimum standard expected by management.

Organizing personnel into individual vehicle maintenance tasks normally results from a number of considerations.

- **Health and Safety — It may be unsafe to have an individual working alone in a particular location or on a particular job. It may, alternatively, be unsafe to have too many personnel working on top of a vehicle simultaneously, or working in a confined space.**

- **Efficient Use of Time — Personnel need to be focused on the task at hand in order to work at a good pace. People in ever larger groups will find more topics to discuss and distract them from their work, so the size of the maintenance team should be kept as small as practical.**

- **Expertise — Different tasks and vehicles can require different sets of expertise and experience. On particular jobs, personnel may need to share their skill sets to overcome particular technical problems.**

Working teams of two personnel is generally effective for routine vehicle maintenance tasks. Depending upon the type of vehicle and the practicality of more personnel sharing the workload simultaneously, the team size can increase if a task is urgent. A single mechanic working alone will often be sufficient for routine tasks such as servicing light vehicles.

In a recessionary climate, there is often a tendency to try and reduce the numbers of maintenance personnel. However, without placing vehicles into long-term storage, or otherwise reducing the fleet size, the effect of reductions in maintenance personnel is a rapid increase in the numbers of vehicles off-road. The recessionary economic situation requires an element of thinking outside the box because it can be difficult or impossible to take on additional permanent staff. Using limited, such as three- or five-year contracts can get around the difficulty of hiring into permanent positions. Recessions also often result in a glut of exceptionally qualified maintenance professionals who are available for contractual work.

A method often used by many businesses to meet various requirements is sub-contracting jobs to other companies. Sub-contracting can be an effective means of overcoming short-term personnel shortages or increases in maintenance demand. It can be used instead of, or in addition to, an overtime facility to meet excess demand on a temporary basis, while maintenance system improvements are achieved. However, it is generally not efficient in the longer term. Outsourcing tasks is most economical for irregular or specialized tasks — those times for which it would be more expensive to employ the required knowledge and skills full time.

For tasks such as component refurbishment (e.g., engine, transmission), the refurbishment and repair of ancillary equipment, or collision repairs, the specialized skills of an external company can be economical. As the tasks become more frequent, the value of outsourcing steadily reduces to a point where it is more economical to move the work in-house. Moving such work in-house becomes more viable because you know your fleet requirements and circumstances better than any external maintenance provider. The move also makes sense because there is sufficient turnover of these tasks to justify any capital equipment expenditure as well as the training and wages of staff to conduct the work.

Sub-contracting work is also suitable for unexpected situations where, for example, worker availability can be adjusted by using external staff to meet temporary increases in demand. A company to which work is being contracted out should generally incur similar costs to the in-house maintenance system, plus their profit margins or fees added on top of this. There is no secret technique involved in external maintenance that cannot be applied in-house.

There are few reasons why in-house costs, in a large well-established maintenance organization, should be exceptionally higher than costs in an external company for a similar quantity and type of work. If they are, then a serious review of costs must be carried out. It may be the case that an outsourced job is cheaper because the external workers are paid less than your own employees. Other costs such as avoiding exposure to vacancy, overtime bills, and health and safety compensation risks, as well as an outside company's ability to order a larger quantity of maintenance materials, can combine to reduce the cost of sub-contracting work. The costs of taking sub-contracted work back in-house also need to account for the direct maintenance costs, such as spare parts and ancillary expenses, that were previously included in the external contractor's fee and now need to be considered in addition to the wages of additional in-house staff. These factors must also be accounted for prior to switching to an in-house system.

In the case of military organizations or those civilian organizations using unique equipment, sub-contracting of work may not be commonplace. The reasons include impracticalities related to maintenance personnel in an external company undergoing training on specialized equipment and vehicles. There can also be difficulties with the provision of unique spare parts, which a subcontractor may not be able to easily source.

For the management of personnel to be effective on a task-by-task basis, there must be measurable standards in place by which to rate the staff. Individual maintenance tasks should be quantified regarding their estimated time for completion, according to a standard method of computation, so that any issues with underperforming maintenance personnel can be accurately identified and addressed. Such quantification is often achieved through a time-study, which can then lead into an analysis of each task. Following this time-study, a work simplification process can be undertaken to improve the maintenance staff's overall performance.

In most western countries, the nature of modern employment legislation requires keeping accurate written records of any issues with staff and the means and timing by which the staff members were informed of each issue with their performance. To accurately quantify underperforming staff who are not completing their tasks at an appropriate rate, you must have a record not only of when they started and completed a task, but also the points at which secondary support tasks were achieved by the organization, such as the supply of spare parts. This information allows management to identify if delays were not caused by the employees directly, but instead were actually out of their control. Recording how long the same task takes other employees to complete gives an additional benchmark measurement. The most accurate method to measure these variables is through the use of a working time clock or scan/swipe system with individual ID cards for each employee.

It is essential that there is a high level of cooperation among all employees, at all levels of the organization, so that both management and direct maintenance staff are working toward the same goals of improved performance and productivity. Any significant divergence in the overarching perspective of different staff towards the place of maintenance in the fleet will introduce difficulties to the maintenance system. All support tasks such as spare parts supply, cleaning, diagnostics, or any other, must understand that they exist solely to support the maintenance function.

The importance of the overall working environment can also often be overlooked in the rush to complete direct tasks. If personnel arriving to work each day enter a dirty, dark, or disorganized environment, they can develop negative attitudes. For example, the importance of the indirect function of cleaning staff in the workplace is key to the attitude of other personnel, and is often underestimated. No different from the old adage that the quality of a hotel can be judged by its bathrooms, the quality of workplace cleanliness tells a similar story.

It can be of benefit to increase each employee's level of understanding of the job immediately above them in the employment structure. This seems to rarely happen in the current economic environment where staff at all levels are afraid of being replaced, reassigned, or let go. Consequently, they become ever more insular and protective of their individual role in the company. Giving employees more understanding of the reasons why their immediate superior is making specific decisions, or the thought process behind certain types of tasks, can give them an appreciation of the bigger picture, resulting in improved morale and work turnover.

Such an aim can be achieved through presentations on the organization's employment structure, or having employees understudy their manager for a period of time. It is preferable that this presentation is made upon their initial employment. Regardless, it requires strong, confident, and competent superiors to give potential successors a good insight into their job.

It is essential for the maximum productivity of personnel that maintenance tasks are planned ahead of time and jobs are queued by the floor level managers to ensure that unnecessary downtime is minimized. Such planning should include all vehicles presently in the maintenance queue, even if it involves planning work for each maintenance team 3 or 4 jobs in advance. This plan, like any other, is a guideline; it can be flexible or changed as required.

The further in advance that such jobs are queued, the more flexible the plan will have to be. However, if there is more than one waiting job queued for each maintenance team, you have inefficiencies in your operational and maintenance setup that should be assessed and rectified. The prior planning of all maintenance tasks will not only facilitate the identification of such backlogs, but also allow a more useful perspective on estimated time to repair — to be given to the operators awaiting return of their vehicles.

Although the spares supply function exists solely to facilitate the maintenance function, the maintenance function exists solely to facilitate the vehicle operators. This relationship is often forgotten by maintenance managers who feel a sense of ownership of the vehicle fleet. A simple whiteboard located on the workshop floor and maintained by the floor supervisor will allow all personnel to know the quantity and type of vehicles awaiting attention. This board can ease the continuation of daily maintenance if the floor supervisor is on leave or otherwise unavailable. Although information is power, the elimination of a gatekeeper mentality at all organizational levels and the sharing of information to improve maintenance efficiency is essential to achieve maintenance excellence.

Insofar as possible, the planning of all queued jobs should include checking for parts availability. Although such a task is not directly related to managing personnel themselves, it does tie into the overall time management of your personnel. Ensuring parts availability now will avoid much wasted time later.

When determining the total quantity of personnel required in your maintenance staff, both the historic data collected regarding vehicle failures and the associated repair times for these failures can be extrapolated to give the total repair time required annually for the fleet. Dividing this number by the working time available for one mechanic during a full year will give an approximation of the number of mechanics required. Prior to completing this calculation, it is important to subtract items such as coffee breaks, washing up time prior to work breaks, restroom breaks, etc., from the mechanic's annual available working time. Also, if the total quantity of employees is slightly below the required annual maintenance hours, it may motivate management to pursue productivity and efficiency improvements, while overtime and outsourcing can temporarily meet any excess. If there are too many employees they will create extra tasks or work slower in order to justify their positions. Hence, it is uneconomical to have excess full-time staff before sufficient additional work is available.

# APPENDIX 1

## VEHICLE TELEMATICS

As with Life Cycle Costs, there are entire books available on the topic of vehicle telematics; therefore, this is intended only as a very short overview.

Modern equipment and the widespread use of electronics in modern vehicles make it possible to monitor and measure almost any input to a particular machine. We can also measure and record the state of a particular piece of equipment or vehicle at any moment in time. This information can be monitored remotely in real time and saved for later access and training usage. This trend will steadily increase in coming years.

**Telematics is the gathering and transmission of information and data to or from a vehicle or piece of equipment.**

Although telematics is the gathering and transmission of information and data, the nature of this information and data can vary depending upon the requirements of the fleet operator. An increasingly common use of telematics is the remote monitoring of cargo doors and global location on the trailer units of high value loads such as ink-jet printer cartridges. Such systems allow fleet operators to monitor, from a central control room, the location of all their vehicles in real-time. For example, they can record when doors are opened or closed to ensure this happens only at designated drop locations. For medium value loads, real-time monitoring may not be deemed necessary; the data can be recorded for use later if cargo is missing or reported lost. Individuals generally have an aversion to being monitored. As such, the introduction of such monitoring systems can meet resistance from staff and unions, requiring a strong argument to gain buy-in.

In addition to security benefits such as those described above, there can also be operational financial benefits derived from a vehicle or driver monitoring system. Such benefits are primarily indirect benefits achieved as a result of modifying driver behavior. For example, if we monitor fuel usage and rpm, there will be no reduction in vehicle fuel usage unless we effectively use this information to educate the drivers to manage the vehicles' rpm when driving and to minimize idling time. If our fleet is large enough to justify the expense of a permanently manned control room, or if we have such a facility due to our security requirements, we can also use real-time telematics for route optimization in the event of traffic delays.

Recording existing fleet costs prior to purchase is a factor often overlooked by organizations considering vehicle monitoring systems to reduce operating costs and optimize routes. The existing costs must be clearly defined before pursuing a telematics purchase in order to fully quantify the savings made after installation of such a system. There is no doubt that the clearest and greatest savings are to be made regarding fuel usage. Savings can result from lower fuel consumption caused by better rpm management as mentioned, or from reduced vehicle speeds as operators become aware that their speed in the event of an accident, or even without being in an accident, can be checked either in real time or after a download.

Monitoring systems can be used to provide evidence of misbehavior, such as speeding by drivers. However, monitoring systems can also be used to help prove a driver's innocence. On numerous occasions, I have generated monitoring system outputs for military commanders regarding data from vehicle and driver monitoring systems, where the data indicated that vehicle drivers were not exceeding speed limits at the time of various events.

It is important to note that, at present, the use of vehicle monitoring systems data recording in civilian or military courts has not been widely proven worldwide; we have yet to see how it will withstand the rigors of legal process in the future. However, many companies do use it to reward or penalize driver behavior based upon their in-house company procedures. Driver monitoring systems can be calibrated to grade drivers based upon speed, harsh braking, gear shifting, and other monitored points. Such grading can be used in many ways, such as to provide an increased end-of-year bonus to the best driver. This practice can motivate all personnel to improve their performance — in turn, reducing our vehicle wear and fuel usage.

When considering a Telematics system, the specific type of components, such as fuel senders and gauges fitted to your vehicles, must be compatible with your intended monitoring system. This and similar issues can be clarified by the intended supplier, but the most accurate method of fuel measurement is to fit the vehicle fuel lines with flow meters and take your readings directly. I have seen monitoring systems where the type of fuel gauges in particular vehicles did not generate a clear enough electrical signal for the modern digital monitoring system to receive and record fuel level accurately. This example is only one of almost infinite incompatibility issues that can arise depending upon your chosen monitoring system and fleet profile.

The extent of a monitoring system installation can vary greatly depending upon individual requirements, but the amount of training required for your organization to self-administer such a system should not be underestimated. If possible, the administration of the system on an on-going basis should be carried out by the supplier; otherwise, you will need one or more employees to take on this task full time. The personnel managing the system must be sufficiently versed in the system to give evidence in court — for example, regarding its accuracy if necessary.

The administration of an installed system by a

supplier is not an unusual occurrence. Most of the established providers operate on this basis. They can download data over the internet at regular intervals and send you regular reports. Or they can send reports on specific events as they arise, or as requested. Having the supplier manage the system, including such tasks as physical repairs or annual calibration, can transfer a lot of the responsibilities for accuracy of the recordings away from your company. Such ongoing management of the system comes at a price, usually in the form of an annual subscription with addition charges for non-standard reports. Check in your own country where vehicle telematics sits in your national legal structure; in some countries, the act of monitoring and recording drivers' actions is itself illegal.

A basic fundamental system for vehicle monitoring will include speed monitoring, usually on a second-by-second basis, and also the monitoring of service brakes. Beyond these two monitored points, the scale of monitoring is limited only by your organization's needs and finances. Items to be monitored can include gear changes, lights, handbrake, door opening/closing, GPS location, rate of acceleration, and many others. It is possible, as previously suggested, to monitor all vehicle locations and speed in real-time through the use of GPS tracking with data being sent back to a monitoring station via mobile phone networks. This does not have to be very expensive; the information packets being sent from the vehicles can be very small and may need to be sent only every 5 minutes or so. By remotely controlling the system settings, the tracking interval for any individual vehicle can be altered by the control room to every few seconds, if there is a on-going event such as a theft or hijacking.

Remote diagnostics is another facet of telematics which is likely to increase in use over coming years and decades. Remote diagnostics can help identify faults with vehicles in the field. These faults can then be remotely rectified, for example, by changing ECU settings for fan operation if a vehicle is in danger of overheating. Remote diagnostics can also be used to tell vehicle operators that a vehicle needs to report to a maintenance location or service depot due to a detected problem, thus avoiding the problem degrading into a larger issue. In addition, remote diagnostics can compile data on the status of a vehicle prior to its arriving for maintenance, either after being recalled or through an automated transmission that is activated when a vehicle drives through the entrance of a service depot. Such actions reduce the diagnostic time required by technicians, giving them an idea of the nature of vehicle faults before the vehicle arrives onto the workshop floor.

Data from the systems can also be downloaded via RF transmission at specified locations such as depots, when vehicles return from their deliveries. In addition, data can be downloaded manually using SD Cards, USB Keys, or the manufacturers' proprietary download devices, if required.

Manual download of data should be avoided at all costs and you should always aim for automatic download using RF or Wi-Fi, or if applicable the automatic transmission of recorded data via the internet.

Downloading multiple vehicles manually can very quickly become a laborious and time-consuming task, causing failures by personnel to download data at the required regular intervals. Ultimately, data will be lost when the in-vehicle memory capacity is full and begins to overwrite itself with newer data.

The use of CAN by a majority of modern vehicles often simplifies the installation of Telematics, although it is also possible to directly wire older vehicles at more cost, if necessary. The use of vehicle telematics can substantially benefit our fleet operations. But the scale of the task involved in its implementation and management should not be underestimated. I strongly recommend contracting the system provider to manage it on an ongoing basis after installation. Failing that, on any reasonably sized fleet, one or more full time employees will be required to manage the system.

Again, the use of any form of manual download is strongly discouraged. RF or Wi-Fi data transfer is essential to these systems operating effectively in a large fleet.

Vehicle telematics is an emerging industry, comprising many different areas and uses as described above; it also includes others not discussed, such as remote entertainment transmission to civilian vehicles. The possible benefits of telematics for fleet operation have the potential to be ever more significant in the future, but the financial savings are indirect. As such, a tangible financial benefit will only be realized if the telematics systems are selected and used in an appropriate manner that fits with your individual fleet operations.

# APPENDIX 2

## KEY POINTS

If you take nothing else away from this book, take these points:

1. The maintenance setup of an organization does not define the On-Road percentage for your fleet. It facilitates the required percentage that is defined by your organization's needs.

2. "Build your team and then trust them to get on with it." (Templer, 2011)

3. Historic analysis (fleet analytics) is how you find the inefficiencies and faults in your system that are costing you money. Record keeping, accurate data, and more record keeping are key.

4. When it comes to significant changes for your fleet, or expenditure related to your fleet, base your decisions only on hard data and facts. If you don't have hard data, refer to Point No. 3 above.

5. Fleet and maintenance data must be maintained digitally and in an easily accessible format. You cannot afford to be tied into a cumbersome or non-functional record keeping system.

6. In the case of new vehicles, infrequent failures from the moment they roll off the production line are key to gaining and maintaining operator confidence. This requires a longer, more thorough, and un-rushed purchasing process.

7. In vehicle telematics, serious consideration should be given to contracting out the ongoing operation of your system. Otherwise, a permanent staff will be required to operate it.

8. In vehicle telematics, the use of any form of manual download is strongly discouraged. RF or Wi-Fi data transfer is essential to these systems operating effectively.

9. A new vehicle can never be all things to all men. Compromises will need to be accepted and made.

10. An RFI can significantly improve the initial stages of any project, at virtually no cost to your organization.

11. Setting timelines on a purchasing project that are too short, or driven by budget dates or organizational politics, is not conducive to value for money or to purchasing the best final product.

12. If your aim is to buy a proven vehicle — one that will enter service as quickly and smoothly as possible — the key is to avoid modifications.

13. Commercial Off The Shelf (COTS) components are the key to low cost, sustainable maintainability; they can significantly contribute to reducing life cycle costs.

14. Buying a fleet of new specialized vehicles is not the same as buying a family car.

15. Ancillary vehicle purchase requirements, such as spares availability, must be assured prior to vehicle delivery.

16. Operator/Driver training must have a direct relevance to the practical operating considerations that will arise with the daily use of the vehicles.

# APPENDIX 3

## COMMONLY REQUIRED CONVERSION FACTORS

**Power**

| | | |
|---|---|---|
| Horsepower → | Kilowatt → | x 0.7459 |
| Kilowatt → | Horsepower → | x 1.3404 |

**Speed**

| | | |
|---|---|---|
| Km/Hr → | Miles/Hr → | x 0.6214 |
| Km/Hr → | Meters/s → | x 0.2778 |
| Km/Hr → | Feet/s → | x 0.9113 |
| Miles/Hr → | Km/Hr → | x 1.6090 |
| Miles/Hr → | Meters/s → | x 0.4470 |
| Miles/Hr → | Feet/s → | x 1.4670 |
| Meters/s → | Km/Hr → | x 3.6000 |
| Meters/s → | Miles/Hr → | x 2.2370 |
| Meters/s → | Feet/s → | x 3.281 |
| Feet/s → | Km/Hr → | x 1.0970 |
| Feet/s → | Miles/Hr → | x 0.6818 |
| Feet/s → | Meters/s → | x 0.3048 |

Acceleration Due to Gravity 9.8067m/s/s

**Volume**

| | | |
|---|---|---|
| Cubic Meter → | Cubic cm → | x 1000000 |
| Cubic Meter → | Liter → | x 1000 |
| Cubic Meter → | U.S. Gallon → | x 264.172 |
| Cubic Meter → | Imperial Gallon → | x 219.969 |
| Cubic Meter → | Cubic Foot → | x 35.3147 |

# Appendix 3

**Volume**

| | | | | |
|---|---|---|---|---|
| Cubic Meter | → | Cubic Inch | → | x 61023.7 |
| | | | | |
| Cubic Centimeter | → | Cubic Meter | → | x 1x10 |
| Cubic Centimeter | → | Liter | → | x 0.0010 |
| Cubic Centimeter | → | U.S. Gallon | → | x 0.000264 |
| Cubic Centimeter | → | Imperial Gallon | → | x 0.000219 |
| Cubic Centimeter | → | Cubic Foot | → | x 0.0000353 |
| Cubic Centimeter | → | Cubic Inch | → | x 0.0610 |
| | | | | |
| Liter | → | Cubic Meter | → | x 0.001 |
| Liter | → | Cubic Centimeter | → | x 1000 |
| Liter | → | U.S. Gallon | → | x 0.2641 |
| Liter | → | Imperial Gallon | → | x 0.2199 |
| Liter | → | Cubic Foot | → | x 0.0353 |
| Liter | → | Cubic Inch | → | x 61.0237 |
| | | | | |
| U.S. Gallon | → | Cubic Meter | → | x 0.00378 |
| U.S. Gallon | → | Cubic Centimeter | → | x 3785.4118 |
| U.S. Gallon | → | Liter | → | x 3.7854 |
| U.S. Gallon | → | Imperial Gallon | → | x 0.8326 |
| U.S. Gallon | → | Cubic Foot | → | x 0.1336 |
| U.S. Gallon | → | Cubic Inch | → | x 231 |
| | | | | |
| Imperial Gallon | → | Cubic Meter | → | x 0.0045 |
| Imperial Gallon | → | Cubic Centimeter | → | x 4546.09 |
| Imperial Gallon | → | Liter | → | x 4.5460 |
| Imperial Gallon | → | U.S. Gallon | → | x 1.2009 |
| Imperial Gallon | → | Cubic Foot | → | x 0.1605 |
| Imperial Gallon | → | Cubic Inch | → | x 277.4194 |
| | | | | |
| Cubic Foot | → | Cubic Meter | → | x 0.0283 |
| Cubic Foot | → | Cubic Centimeter | → | x 28316.846 |
| Cubic Foot | → | Liter | → | x 28.3168 |
| Cubic Foot | → | U.S. Gallon | → | x 7.4805 |
| Cubic Foot | → | Imperial Gallon | → | x 6.2288 |
| Cubic Foot | → | Cubic Inch | → | x 1728 |
| | | | | |
| Cubic Inch | → | Cubic Meter | → | x0.0000163 |
| Cubic Inch | → | Cubic Centimeter | → | x 16.3870 |
| Cubic Inch | → | Liter | → | x 0.0163 |
| Cubic Inch | → | U.S. Gallon | → | x 0.0043 |
| Cubic Inch | → | Imperial Gallon | → | x 0.0036 |
| Cubic Inch | → | Cubic Foot | → | x 0.000578 |

# GLOSSARY

**Addition** — Any alteration to the baseline vehicle that does not change the fundamental engineered properties of its mechanical vehicle components, usually in the form of add-on or ancillary equipment. Including mountings for ancillary equipment or crew kit.

**Area of Responsibility** — A geographical region within a defined boundary for which a military commander, manager, or other person takes responsibility.

**Availability %** — The percentage of a vehicle fleet fully operational and available for use at any given point in time.

**Availability Driven Maintenance** — A maintenance system where the maintenance capacity and turnover are being determined by a lack of ready availability of equipment or personnel resources.

**Backlog** — See *Maintenance Backlog*

**Benefit-Cost Analysis** — A process of evaluating the benefit of something with respect to its complete implementation cost.

**Best Practice** — When there is more than one method of achieving an aim, the method which it is believed will achieve the aim in a better fashion than any other method.

**Breakdown Rate** — The rate or frequency at which vehicles go Off-Road or equipment becomes unserviceable.

## Glossary

**Bureaucracy** — A chain of authority involving many levels of seniority, with many administrators and officials at each level.

**Cannibalization** — A process of removing components from a vehicle in order to fit them during maintenance to another vehicle, usually of the same type.

**Catastrophic Failure** — The complete breakdown of a vehicle or machine from which it cannot be recovered, requiring mechanical repairs to return the unit to operation.

**Class Code** — A designation or number used to indicate an individual type or class of vehicle.

**Co-Location** — When a number of facilities used to carry out different maintenance functions, are located on the same site.

**Component** — A mechanical part of a whole vehicle.

**Contract Maintenance** — A process of paying an external company to conduct vehicle maintenance instead of conducting it within your own organization. This process can be for an entire fleet or on an *as necessary* basis for individual vehicles.

**Cost-Benefit Analysis** — See *Benefit-Cost Analysis*

**Crew Maintenance** — Minor vehicle maintenance tasks carried out by the vehicle operators. Usually involving tasks that do not require the removal and replacement of components.

**Critical Cost Issues** — Tasks, issues or procedures that incur the most significant costs for your organization.

**Depreciable Life** — The total period of time during which your vehicle or equipment holds a financial value above its residual value.

**Depreciation** — The financial value that a vehicle or piece of equipment loses over time after it is purchased.

# Glossary

**Design for Maintenance** — A process of designing vehicles and equipment in such a manner as to ensure that maintenance can be carried out in the easiest and quickest ways possible.

**Design for Reliability** — A process of designing vehicles and equipment in such a manner as to minimize component failures during different types of envisioned usage.

**Devolved budget** — A budget that allows lower level managers to determine how the money in the budget is going to be spent.

**Direct Running Costs** — The financial expenditure required to enable a vehicle to physically travel from one point or location to another.

**Direct Tasks** — Those tasks that physically contribute to repairing or rectifying a fault on a vehicle or piece of equipment. Direct tasks can be delayed by the time taken for preceding indirect tasks.

**Downtime** — The length of time when a vehicle or piece of equipment is non-operational or not capable of fulfilling its required purpose.

**Execution Ratio** — The proportion of jobs fully completed as per the maintenance schedule, relative to the proportion of jobs which that have not been completed in accordance with the maintenance schedule.

**Failed Repair** — A repair that is unsuccessful in returning a vehicle to service for the anticipated length of time prior to a repeat failure.

**Failure Analysis** — A process of investigating failures to determine when, how, and why they have occurred; in order to devise prevention measures to reduce future failures.

**Failure Distribution** — The time or mileage interval at which failures are occurring for a single component type, and the quantity of those failures which are occurring at each interval.

**Failure Frequency of Occurrence** — The number of failures of a single type occurring during a specified time period and the length of time between those failures.

**Failure Rate** — See *Failure Frequency of Occurrence*

**Forced Deterioration** — Deterioration of a vehicle caused by its use in a manner other than its design specification.

**Gatekeepers** — Employees within an organization who withhold information in order to use it for their own gain at a later date or to reinforce their own sense of self-importance.

**Hands On Tool Time** — Time that a maintainer spends physically carrying out maintenance tasks on a vehicle.

**In-Direct Running Costs** — The financial expenditure required in the background to keep a vehicle in such a condition whereby it can incur direct running costs.

**In-Direct Tasks** — Those tasks that precede direct maintenance tasks but are not included in the hands-on repair or resolution of vehicle faults. In-Direct tasks for example include scheduling, recovery, fault diagnosis, and spare parts supply.

**In-House Inventory** — Vehicle spare parts that are being held in stock within an organization, ready to be issued for use as required.

**In-House Maintenance** — Maintenance which is being conducted by employees of the same organization operating the vehicle or vehicles.

**In-Service Modifications** — Permanent modifications implemented by the customer to a vehicle platform after it has already entered service.

**Integrity Priority Tasks** — Maintenance tasks that are carried out to preserve the integrity of a vehicle, such as to prevent future degradation and worse failures of sub-components.

**Life Cycle Costs** — Financial costs which that will be incurred throughout the entire lifetime of a new vehicle.

**Life Extension Program** — Permanent modifications implemented by the customer to a vehicle platform after it has already entered service, in order to extend its useful in-service lifespan.

**LTTA (Lead Time To Availability)** — The amount of time required between ordering a component and its becoming available to the front-line maintainer for usage.

**Maintainer** — A professionally qualified person employed to conduct maintenance and repair tasks on vehicles and their equipment. Includes various discipline such as mechanics, auto electricians, communications technicians, and ordnance technicians.

**Maintenance** — A process of preventing or resolving vehicle or equipment faults and failures.

**Maintenance Backlog** — The queue of Off-Road vehicles that are not yet undergoing hands-on maintenance or receiving the attention of maintainers.

**Maintenance Capacity** — The total quantity of vehicles that can receive the attention of maintainers at a given maintenance location, at any single point in time.

**Maintenance Cycle** — The level of demand for vehicle maintenance experienced by a maintenance system throughout different times of a typical year.

**Maintenance Intensive Period** — A period of time when extraordinary measures are put in place to increase the maintenance capacity and turnover of a given maintenance location.

**Maintenance Performance** — How effectively an overall maintenance system is at meeting the required aims of repairing and returning vehicles to service.

**Maintenance Management System** — A system of scheduling maintenance tasks, tracking their implementation, planning subsequent tasks, and reviewing all maintenance previously carried out.

**Maintenance Scheduling** — A system of pre-planning maintenance tasks, the time they will take place, and an estimate of their time for completion.

**Maintenance System** — The overall process in place to prevent vehicle failures, and return vehicles or equipment to service from the time they begin to malfunction to the time they are again fully functional and operational.

**Malfunctioning** — A vehicle or piece of equipment that is functioning incorrectly or not functioning at all, in accordance with its specifications and design performance.

**Manufacturing Tasks** — Tasks involved in the production of assemblies and components from raw materials.

**Mid Life Upgrades** — Permanent modifications carried out by the customer during the life cycle of a vehicle or piece of equipment, in order to bring the performance up to modern levels or to generally improve performance.

**MMBF (Mean Miles Between Failure)** — The average distance, usually measured in miles or kilometers, between failures of vehicles of the same type.

**Mobile Workshops** – A vehicle repair workshop built into a truck or other vehicle, carrying equipment and spare parts, which that can then travel with groups of vehicles to provide maintenance support away from base.

**Modification** — Any alteration to a baseline vehicle that changes the fundamental engineered properties of its mechanical vehicle components, including removal of moving components and/or their replacement with alternatives.

**MTBF (Mean Time Between Failure)** — The average time, usually measured in operating hours, between failures of repairable components of the same type.

**MTTF (Mean Time To Failure)** — The average time, usually measured in operating hours, between failures of non-repairable components of the same type.

**Off-Road** — A vehicle that is not functional or not operating.

**Operational Fleet** — A fleet of vehicles being used within a geographical operational area.

**Operational Theatre** — A geographical area within a larger theatre of war, where operations are actively taking place.

**Operators** — The end users of a vehicle who are involved in physically operating it on a daily basis.

**Outsourcing** — Using an external supplier or company to provide services or work.

**Post Delivery Inspection** — A vehicle inspection conducted after delivery to the customer, to ensure that a new vehicle has not suffered any damage in transit.

**Pre Delivery Inspection** — A vehicle inspection conducted after production and before delivery to the customer, to ensure that a new vehicle meets the purchase specification and the required quality standards.

**Predictive Maintenance System** — A system of replacing vehicle components prior to failure in order to prevent the failure occurring. The replacement schedule is based upon the observed degradation rate of components in operation.

**Preventive Maintenance System** — A system of replacing vehicle components prior to failure in order to prevent the failure occurring. The replacement schedule is based upon the observed time interval of past failures for each component.

**Process Optimization** — A system of reducing inefficiencies, to result in the minimum time for a process or series of tasks to be successfully completed.

**Production Line** — A sequence of operations within a factory that produces vehicles through a number of consecutive stages.

**Queued Maintenance** — Vehicles or equipment requiring maintenance tasks, that are in a queue awaiting availability of maintainers or other resources before work is initiated.

**Reactive Maintenance** — Replacing or repairing vehicle components in response to a failure that has already occurred.

**Recovery** — The process of collecting an Off-Road vehicle and moving it to a location where it can be repaired.

**Reliability%** — See *Availability*

**Reliability Centered Maintenance** — A process defining the overall maintenance system, used to establish the safe minimum levels of maintenance while still maintaining reliability of the vehicles or equipment.

**Repairables** — Vehicle components that can undergo repairs to return them to use.

**Repair Codes** — A designation or number used to identify the component groups, assemblies, and sub-assemblies that are being repaired on a vehicle platform.

**Replacement Frequency** — The time between scheduled replacements of vehicle components in a preventive maintenance system.

**Replacement Tasks** — Maintenance tasks involving the physical removal and replacement of components on a vehicle platform.

**Residual Value** — The remaining value a vehicle holds at the end of its useful life due to the raw materials present in its construction, or its potential remaining utility for use by a different organisation or in a different role.

## Glossary

**Resource Allocation** — A process of allocating equipment and maintenance resources to particular maintainers to facilitate their tasks being completed as per the maintenance schedule.

**Root Cause** — The underlying original cause that ultimately generated a later failure.

**Root Cause Analysis** — A process of identifying the underlying cause of a failure.

**Safety Critical Tasks** — Maintenance tasks that are carried out to prevent a likely accident or incident occurring due to component failure while the vehicle is in operation.

**Safe Failure Mode** — A vehicle component or assembly that fails in a safe way so as to prevent accident, injury, or damage; for example, a braking system that can only fail with the brakes engaged.

**Scheduling** — See *Maintenance Scheduling*

**Self-Sufficient** — Able to meet all maintenance and operational needs without support from other vehicles.

**Specializations** — Specific areas of vehicle maintenance other than general mechanical maintenance, such as electrical, ordnance, or communications.

**Specialized Vehicle** — A vehicle that is designed and used to fulfil a very specific out-of-the-ordinary role.

**Spurious Parts** — Vehicle or equipment components which are manufactured by an alternative to the original genuine parts manufacturer. Spurious parts may be manufactured to a higher or lower quality standard. In some cases, they may be less expensive but less reliable than genuine parts.

**Standby Vehicle** — A fully serviceable vehicle not allocated for use that can be rushed into use when required to replace a vehicle which has gone Off-Road.

**Stranded Costs** — Costs that cannot be recovered except over the full service life of a vehicle. An investment in new vehicle housing for new vehicles would usually be a stranded cost.

**Time Study** — A process of examining the time taken for each individual element of a task, in order to determine the fastest and most efficient way to complete that task.

**Total Productive Maintenance** — A process of maximizing vehicle On-Road time by minimizing deterioration of maintenance issues and unplanned stoppages of equipment or failures of components.

**Trickle Charging** — Also known as Maintenance Charging or Smart Charging. Charging a vehicle battery at a rate slow enough rate to maintain a full charge state without over-charging, preferably using intermittent current once a battery is fully charged.

**True Maintenance** — Vehicle direct maintenance tasks carried out in order to rectify or prevent a vehicle fault. Usually involving the removal, repair, and/or replacement of components.

**Turnover** — The quantity of vehicles entering and leaving the maintenance system at any given point in time or over a given time period.

**Ultra-Reliability** — The occurrence of no failures, or very few, during the entire usable life of a vehicle or piece of equipment.

**Unallocated Hours** — Time within a maintenance schedule that is not allotted to any tasks in order to allow flexibility in the schedule.

**Unit Cost** — The price of an individual vehicle, component, assembly, or piece of equipment.

**Unplanned Failure** — The failure of a vehicle or its components that was not foreseen by the maintenance schedule.

**Unplanned Maintenance** — See *Reactive Maintenance*

**Unscheduled Maintenance** — See *Reactive Maintenance*

## Glossary

**Unserviceable** — See *Off-Road*

**Upgrade** — An addition or modification to a vehicle which will either improve the mechanical properties of a component, or which will improve the performance of the vehicle related to operations and/or maintenance.

**Usage Profile** — The way a vehicle is being operated, considering all aspects, such as amount of time on-road versus off-road, loads being carried, average time spent at different speeds, etc.

**Variants** — Different versions of the same vehicle that are equipped to fulfil specific and differing roles.

**Vehicle Acceptance** — See *Post Delivery Inspection*

**Vehicle Baseline** — The design of a vehicle as per its original specifications and engineering standards prior to upgrades, modifications, changes, or redesign.

**Vigilance Decrement** — Also known as cognitive fatigue. The inability of a person to remain vigilant for errors, a steady reduction in their vigilance, or a slowing in their reaction time.

**Warranty Modifications** — Permanent modifications carried out by the manufacturer on a new vehicle after delivery, free of charge, to rectify a production fault or a design error.

**Weakness Analysis** — Analyzing a process or series of tasks to determine where failures are likely to occur or efficiency losses are being incurred in a system.

# REFERENCES


2009. *Collins English Dictionary — Complete & Unabridged*, 10th Edition. Collins.

2012. *Average Automotive Mechanic I Salary Information plus Job, Career Education & Unemployment Help*. [ONLINE] Available at: http://www1.salary.com/Automotive-Mechanic-I-Salary.html.

2012. *The Free Dictionary*. Farlex. [ONLINE] Available at: http://www.thefreedictionary.com.

Anderson, Lee G. and Russell F. Settle. 1977. *Benefit-Cost Analysis: A Practical Guide*. D. C. Heath and Company.

Astphan, Alyssa A. October 2005. *FA Journal: Two Level Maintenance Modularity and the Transformation of Army Maintenance.*

Borris, Steven. 2006. *Total Productive Maintenance*. McGraw-Hill.

Cork County Council, Ireland-Cork: Fleet management, repair and maintenance services - Ref: 2014/S 194-342980

Corporate Manslaughter. *Legal Guidance: The Crown Prosecution Service*. 2014. [Online] Available at: http://www.cps.gov.uk/legal/a_to_c/corporate_manslaughter/

Cunningham, Dominic. April 1995. *Án Cosantoir: Future Trends in Vehicle Maintenance.*

Department of Defence Ireland, 'Competition for the Supply of Light Tactical Armoured Vehicles (LTAVs) - Ref: Con/008/2008', 08th May 2008, [Online] Available at: http://www.defence.ie

# References


Dolce, John E. 2009. *Analytical Fleet Maintenance Management, 3rd Edition*. SAE International.

Dublin Bus, 'Supply of Support Vehicles - Ref: 2014.169'

Fan, Carol E., Eric Peltz, and Lisa Colabella. 2005. *The Effects of Equipment Age on Spare Parts Costs: A Study of M1 Tanks*. RAND Corporation.

Fanning, Dean, CCE, PE. 2014. *Life Cycle Costing: How to calculate life cycle costs and total ownership Cost*. CreateSpace Independent Publishing Platform.

Foy, Dennis. 2002. *Automotive Telematics: The One-Stop Guide to In-Vehicle Telematics and Infotainment Technology and Applications*. Red Hat Publishing.

Friedli, Lucas. 2011. *Repairing the Panzers Volume II*. Panzerwrecks.

Irish Coast Guard, '4-wheel-drive-vehicles - Ref: 2014/S 184-324369'

Irish Coast Guard, 'Ireland-Dublin: Trailers, semi-trailers and mobile containers - Ref: 2014/S 184-324437'

Lindenmeyer, Vincent and Gilbert Duran. January–February 2001. *Army Logistician Professional Bulletin of United States Army Logistics: Reducing Maintenance Backlog*.

Linn, Mike C. J. and Walter P. Smith. 1999. *Process Optimization Guide for Military Manufacturing and Maintenance Facilities*. U.S. Army Corps of Engineers.

Mobley, R. Keith. 2008. *Maintenance Engineering Handbook,* 7th Edition. McGraw Hill.

Moubray, John. 2007. *Reliability-Centred Maintenance, 2nd Edition.* Elsevier, Ltd.

Narayan, V. 2012. *Effective Maintenance Management Risk and Reliability Strategies for Optimizing Performance,* 2nd Edition. Industrial Press.

Newton, Richard. 2010. *Brilliant Checklists for Project Managers: Your Shortcut to Success,* 2nd Revised Edition. Pearson Education.

Office of Government Procurement, 'Ireland-Dublin: All Terrain Vehicles - Ref: 2014/S 192-338691'

# References


Payant, Richard P. and Bernard T. Lewis. 2007. *Facility Managers Maintenance Handbook,* 2nd Edition. McGraw-Hill.

Price, Richard. September–October 1999. *Army Logistician Professional Bulletin of United States Army Logistics: Ultrareliability Pillar of the AAN.*

Reason, James and Alan Hobbs. 2003. *Managing Maintenance Error.* Ashgate.

Salary.com Salary Wizard- Do you know what you're worth? . 2014. *Salary.com Salary Wizard- Do you know what you're worth?* . [ONLINE] Available at: http://swz.salary.com/salarywizard/

Scorpion Shoot Glen of Immal (8) | Flickr - Photo Sharing!. 2014. *Scorpion Shoot Glen of Immal (8) | Flickr - Photo Sharing!.* [ONLINE] Available at: https://www.flickr.com/photos/dfmagazine/

Stevenson, Mitchell. September–October 2002. *Army Logistician Professional Bulletin of United States Army Logistics: Army Maintenance Transformation.*

Templar, Richard. 2011. *Rules of Management: A Definitive Code for Managerial Success,* 2nd Edition. Prentice Hall.

# INDEX

# Index

# Index